Brain Drain and Gender Inequality in Turkey

Adem Yavuz Elveren

Brain Drain and Gender Inequality in Turkey

Adem Yavuz Elveren
Fitchburg State University
Fitchburg, MA, USA

ISBN 978-3-319-90859-5   ISBN 978-3-319-90860-1 (eBook)
https://doi.org/10.1007/978-3-319-90860-1

Library of Congress Control Number: 2018944118

© The Editor(s) (if applicable) and The Author(s), under exclusive license to Springer International Publishing AG, part of Springer Nature 2018
This work is subject to copyright. All rights are solely and exclusively licensed by the Publisher, whether the whole or part of the material is concerned, specifically the rights of translation, reprinting, reuse of illustrations, recitation, broadcasting, reproduction on microfilms or in any other physical way, and transmission or information storage and retrieval, electronic adaptation, computer software, or by similar or dissimilar methodology now known or hereafter developed.
The use of general descriptive names, registered names, trademarks, service marks, etc. in this publication does not imply, even in the absence of a specific statement, that such names are exempt from the relevant protective laws and regulations and therefore free for general use.
The publisher, the authors and the editors are safe to assume that the advice and information in this book are believed to be true and accurate at the date of publication. Neither the publisher nor the authors or the editors give a warranty, express or implied, with respect to the material contained herein or for any errors or omissions that may have been made. The publisher remains neutral with regard to jurisdictional claims in published maps and institutional affiliations.

Cover illustration: Abstract Bricks and Shadows © Stephen Bonk/Fotolia.co.uk

Printed on acid-free paper

This Palgrave Pivot imprint is published by the registered company Springer International Publishing AG part of Springer Nature.
The registered company address is: Gewerbestrasse 11, 6330 Cham, Switzerland

To my elder sister, Süheyla İnan

ACKNOWLEDGMENTS

I am indebted to Gülay Toksöz for her supervision, to Ünal Töngür for his constant help, and to my daughter, Asya, for picking the cover design. I appreciate the generous support of the Presidency for Turks Abroad and Related Communities, Republic of Turkey Prime Ministry on the brain drain survey. However, needless to say that the opinions expressed in this book are my own and do not reflect the view of the institution.

Contents

1	Introduction	1
2	Brain Drain: Causes and Consequences	7
3	Brain Drain in Turkey: A Literature Review	23
4	A Brain Drain Survey	63
5	The Transformation of Regime and Gender Inequality in Turkey	109
6	Conclusion and Policy Recommendations	131
	Appendix	137
	Index	169

List of Abbreviations

AKP	Justice and Development Party (Adalet ve Kalkınma Partisi)
BİDEB	TÜBİTAK Directorate of Science Fellowships and Grant Programmes (Bilimsel İnsanı Destekleme Daire Başkanlığı)
CCT	Conditional Cash Transfer Program
DPT	State Planning Organization (Devlet Planlama Teşkilatı)
FDI	Foreign Direct Investment
GHI	General Health Insurance
HDC	Highly Developed Country
ILO	International Labour Organization
IMF	International Monetary Fund
İŞKUR	Employment Agency (İş ve İşçi Bulma Kurumu)
LDC	Least Developed Country
LFP	Labor Force Participation
MCA	Multiple Correspondence Analysis
MoE	Ministry of Education
NGO	Nongovernmental Organization
OECD	Organisation for Economic Co-operation and Development
PKK	Kurdistan Workers' Party
RP	Welfare Party (Refah Partisi)
TÜBİTAK	The Scientific and Technological Research Council of Turkey (Türkiye Bilimsel ve Teknolojik Araştırma Kurumu)
UNESCO	United Nations Educational, Scientific and Cultural Organization

LIST OF FIGURES

Fig. 4.1	Current return intention—bachelor degree (all sample)	77
Fig. 4.2	Current return intention—bachelor degree (students)	77
Fig. 4.3	Current return intention—bachelor degree (professionals)	78
Fig. 4.4	Current return intention—bachelor degree (male)	78
Fig. 4.5	Current return intention—bachelor degree (female)	79
Fig. 4.6	Initial return intention—bachelor degree (all sample)	81
Fig. 4.7	Initial return intention—bachelor degree (students)	81
Fig. 4.8	Initial return intention—bachelor degree (professionals)	82
Fig. 4.9	Initial return intention—bachelor degree (male)	82
Fig. 4.10	Initial return intention—bachelor degree (female)	83
Fig. 4.11	Current return intention—field of study (all sample)	85
Fig. 4.12	Current return intention—field of study (students)	85
Fig. 4.13	Current return intention—field of study (professionals)	86
Fig. 4.14	Current return intention—field of study (male)	86
Fig. 4.15	Current return intention—field of study (female)	87
Fig. 4.16	Initial return intention—field of study (all sample)	89
Fig. 4.17	Initial return intention—field of study (students)	89
Fig. 4.18	Initial return intention—field of study (professionals)	90
Fig. 4.19	Initial return intention—field of study (male)	90
Fig. 4.20	Initial return intention—field of study (female)	91
Fig. 4.21	Initial-current return intentions—stay duration (all sample)	101
Fig. 4.22	Initial-current return intentions—stay duration (students)	101
Fig. 4.23	Initial-current return intentions—stay duration (professionals)	102
Fig. 4.24	Initial-current return intentions—stay duration (male)	102
Fig. 4.25	Initial-current return intentions—stay duration (female)	103

LIST OF TABLES

Table 3.1	Empirical literature on the brain drain in Turkey	30
Table 4.1	Age and gender (%)	64
Table 4.2	Country of highest degree	65
Table 4.3	The most important reason for going abroad (%)	67
Table 4.4	The most important reason for going abroad (%) (students + professionals)	67
Table 4.5	Return reasons (%)	68
Table 4.6	Plan to go abroad again and gender (%)	69
Table 4.7	Plan to go abroad again and gender (%) (students + professionals)	69
Table 4.8	Marital status and the status of spouse	70
Table 4.9	Effect of 2008–2009 economic crisis in the US on views about returning	70
Table 4.10	Initial-current comparison, (%) students	71
Table 4.11	Initial-current comparison, (%) professionals	72
Table 4.12	Initial return intention, current return intention	72
Table 4.13	Initial non-return intention, stay duration	72
Table 4.14	Current non-return intention, stay duration	72
Table 4.15	Initial non-return intention, age group	73
Table 4.16	Current non-return intention, age group	73
Table 4.17	Current return—highest degree country, (%) students	73
Table 4.18	Current return—highest degree country, (%) professionals	73
Table 4.19	Current return—highest degree country (%) (students + professionals)	74
Table 4.20	Current non-return intention, highest degree	74
Table 4.21	Initial return—highest degree country, (%) students	74
Table 4.22	Initial return—highest degree country, (%) professionals	74

Table 4.23	Initial return—highest degree country (%) (students + professionals)	75
Table 4.24	Initial non-return intention, highest degree country	75
Table 4.25	Current return—bachelor degree discipline, (%) students	76
Table 4.26	Current return—bachelor degree discipline, (%) professionals	76
Table 4.27	Current return—bachelor degree discipline (%) (students + professionals)	76
Table 4.28	Initial return—bachelor degree discipline, (%) students	80
Table 4.29	Initial return—bachelor degree discipline, (%) professionals	80
Table 4.30	Initial return—bachelor degree discipline (%) (students + professionals)	80
Table 4.31	Current return—field of study, (%) students	83
Table 4.32	Current return—field of study/work, (%) professionals	84
Table 4.33	Current return—field of study/work (%) (students + professionals)	84
Table 4.34	Initial return—field of study, (%) students	87
Table 4.35	Initial return—field of study, (%) professionals	88
Table 4.36	Initial return—field of study (%) (students + professionals)	88
Table 4.37	Current return—having previous experience (study, work, travel) abroad, (%) students	91
Table 4.38	Current return—having previous experience (study, work, travel) abroad, (%) professionals	92
Table 4.39	Current return—having previous experience (study, work, travel) abroad (%) (students + professionals)	92
Table 4.40	Initial return—having previous experience (study, work, travel) abroad, (%) students	92
Table 4.41	Initial return—having previous experience (study, work, travel) abroad, (%) professionals	93
Table 4.42	Initial return—having previous experience (study, work, travel) abroad (%) (students + professionals)	93
Table 4.43	Push and pull factors viewed as important (%) bachelor degree discipline	94
Table 4.44	Push and pull factors viewed as important (%) bachelor degree discipline (students + professionals)	96
Table 4.45	Push and pull factors viewed as important (%) field of study	97
Table 4.46	Push and pull factors viewed as important (%) field of study (students + professionals)	98
Table 4.47	Push and pull factors viewed as important (%) having previous experience (study, work, travel) abroad	99
Table 4.48	Push and pull factors viewed as important (%) having previous experience (study, work, travel) abroad (students + professionals)	100

Table 4.49	Push and pull factors viewed as important (%)	103
Table 4.50	Push and pull factors rated as 'Important' or 'Very Important' (%) (students + professionals)	105
Table 4.51	Non-return intention, regression results	106

CHAPTER 1

Introduction

Abstract The secular and democratic character of the regime in Turkey has eroded under the current ruling Justice and Development Party in the last decade. Islamic values have gradually come to dominate every aspect of life, and women's subordinate roles have been reinforced. As the regime becomes more oppressive more people from the secular side of the country have fled, including the well-to-do, minorities, and academics. Based on an original survey, the book comprehensively analyzes this brain drain from Turkey, focusing on its gender dimension. The book provides a brief account of the transformation of (welfare) regime to show that women's higher tendency not to return to Turkey is a mirror reflection of the gender gap in the labor market and increasing Islamic authoritarianism.

Keywords Brain drain • Gender • High-skilled migration • The Justice and Development Party • Authoritarianism

The West's admiration for the Justice and Development Party (AKP) in the early 2000s claimed that Turkey had finally overcome the enduring Islamist versus secularist societal split and that the Muslim democratic AKP was building a liberal, pluralist society, and this 'New Turkey' could be a role model for the Arab World. However, history proved that Turkey's secular elite was correct as they argued that the AKP or Erdoğan have a hidden agenda to replace Kemalist ideology with an authoritarian Islamic one.

This regime change has led to a great exodus. As the religious nationalist regime becomes more repressive, more people, basically from the secular side, fled Turkey, including the well-to-do, minorities such as Turks of Jewish descent, and academics. This book is about this migration of high-skilled labor, namely, brain drain, from Turkey. It underscores the gender dimension of the brain drain to show that women's higher tendency not to return to Turkey is a mirror reflection of the gender gap in the labor market and increasing Islamic authoritarianism during the current ruling party. To this end, the book analyzes brain drain based on an original survey and provides a brief account of the transformation of (welfare) regime in the era of AKP.

This book aims to examine the brain drain in Turkey, focusing on the determinants of the return intentions of the students and professionals. Another objective of the study is to review the brain drain literature in Turkey comprehensively. One specific goal of the study is to analyze the gender dimension of the brain drain with Turkey, an issue that has not received enough attention. Two different comprehensive questionnaires (46 questions for professionals and 53 questions for students) were prepared for the Turkish students and professionals living abroad. The links to online questionnaires were distributed through the social media and were emailed to relevant social and professional email groups for about five months between late 2015 and early 2016. The study is based on the data derived from that survey, in which 116 students and 84 professionals responded to all the questions.

What is the brain drain? Brain drain is the migration of highly educated individuals from their home countries to developed countries to reach greater opportunities in their field of specialization and/or to have better living conditions and lifestyle.

Why does it matter? Brain drain is considered one of the most detrimental aspects of international migration from the viewpoint of the home country. It is simply the 'reverse technology transfer', a cheap means for developed countries to acquire high-skilled labor. Absent such a group of highly educated individuals, the developing countries can suffer from the lack of a spillover effect. That is to say, in the absence of highly skilled persons, the information and knowledge cannot spread in the society, and there will be fewer people to demand better governance and civic society, which together reduce the productivity in the economy. Also, countries that lost their highly skilled citizens experience difficulty in the delivery of an adequate level of health and education at the desired quality level, which deteriorates the long-term growth potential of the country. It is also argued that the brain drain can benefit the sending country, particularly if the country can generate 'skill mobility' and build the strong and effective ties with their citizen scholars abroad. This is particularly important for such a large country like Turkey because it has long been one of the top ten countries that send their students to pursue graduate degrees in the US. Turkey ranks fourth according to per capita doctoral recipients from the US universities in the years between 2004 and 2014. Turkey, with its large population, has an enormous human capital potential, which is the key element of economic growth. Therefore, it is important to establish an effective education, science, and technology policy. The brain drain is one key issue that must be addressed in such an action plan. Therefore, this research is a modest attempt to provide some more empirical evidence on the subject matter to help policymakers to set up the long-term development strategy.

What do we know about the brain drain in Turkey? Although there exists much literature on migration in Turkey, whose significant part focuses on emigrants in Europe, particularly in Germany, there are fewer works on the brain drain and even fewer empirical ones. There are only a few comprehensive empirical investigations. This is one main motivation for this study, to provide more useful information to the brain drain literature in Turkey.

The previous empirical works on the brain drain in Turkey provided useful information. First, that 'academia/science (academicians/scientists) in Turkey does not receive the attention/appreciation it deserves' is one common argument reported in several studies from the first empirical study on the topic in 1968 to today.

Second, there have been consistent criticisms about the current academic environment in Turkey and concerns about the future of education and science in the country by scholars who received their graduate degrees from a foreign institution and returned to work in Turkey. Accordingly, inadequate library facilities, devoting insufficient funds for scientific activities, deprived conditions of labs in the fields of natural sciences, excessive teaching load for academics, and the absence of a long-term plan for science and technology are some major criticisms often reported. The frequently cited suggestions to improve the academic environment and to prevent brain drain are to extend academic freedom, to increase the number of independent research institutes, and to separate the research and teaching universities. In this context, it is shown that the implementation of the compulsory service[1] is one effective means to prevent the brain drain.

Third, only one study, which was conducted immediately after the 2001 economic crises in Turkey, reported that the economic crises in Turkey are a major factor that affects the return decision of students/scholars abroad (Güngör 2003). However, all other studies showed that the economic situation in Turkey is not a factor that has a significant influence on the return decision. Some studies argued that the recent economic improvement (with some financial incentives) caused a reverse brain drain. However, this positive pattern has been diverted by the worsening situation in the political arena, namely, political instability in the last few years. In other words, 'economic crises' or 'lower pay' in Turkey have not been reported as major reasons for not returning. In fact, some respondents reported that they do not think that their living standards would be lower in Turkey as the pay level in some top private universities is highly satisfactory. That means that, as some studies emphasize, political instability and bureaucratic obstacles are more important concerns than economic factors.

Fourth, it is also shown that the harmonization of policies with the European Union (EU) would have a positive effect on the return decision; there are significant positive correlations between stay time abroad, having a degree from a foreign university/or a university whose medium of language is English, and having previous work/study/travel experience in abroad and the decision to not return; and based on a study by Esen (2014) the 2008–2009 economic crisis in the US appears to have no impact on the return decisions of the Turkish scholars/students.[2] Finally, some studies found that females are more likely to go abroad and not return (Güngör and Tansel 2008, 2014).

What does this study contribute to the literature? This study supports most of the above findings and provides some updated and different information. The effectiveness of compulsory service rule; insignificance of the economic condition in Turkey on the return decision; the political instability being the key factor on return decision, the positive relationship between stay time abroad and not returning, having a degree from a foreign institution/or a university whose medium of language is English, and previous work/travel/study experience abroad and tendency not to return; no effect of the economic crisis in the US on the return decision; and higher tendency to migrate/stay abroad for females are supported by this study.

The study found that aforementioned findings are still valid today, and some of those factors have become more remarkable. For instance, political instability as a push factor becomes more dominant compared to other push (and pull) factors. Also, as a result of the changing paradigm toward the role of women in the society (increasing pressure on women in the social sphere and trying to confine women in the private sphere), women have a higher tendency than men to migrate and not to return. The paper also found that female students are more likely to receive support from their families to go abroad. It is found that women have a higher tendency to go abroad and prefer to stay longer mainly because of push factors than because of pull factors.

The outline of the book is as follows: Following this chapter, the causes and consequences of the brain drain are discussed in Chap. 2. Chapter 3 comprehensively discusses the brain drain in Turkey. In Sect. 3.1 in Chap. 3, nonempirical literature on the brain drain is presented. Section 3.2 in Chap. 3 focuses on the findings of the empirical works. Section 3.3 in Chap. 3 deals with the current debates and the reverse brain drain. Chapter 4 presents the findings of the survey. Chapter 5 provides a brief account of the transformation of the regime and gender inequality in Turkey. Finally, the concluding chapter (Chap. 6) is reserved to summarize the findings and provide some policy recommendations.

Notes

1. Compulsory service refers to required service in a preassigned peripheral university of students who receive a scholarship from the government to pursue their graduate degrees.
2. Throughout the study 'Turkish' refers to the people of Turkey (i.e., citizens of Turkey) and does not refer to ethnicity by any means.

References

Esen, E. (2014). *Going and Coming: Why U.S.-Educated Turkish PhD Holders Stay in the U.S. or Return to Turkey?* Unpublished PhD dissertation, Department of Educational Leadership and Policy Studies, The University of Kansas, Lawrence, KS, USA.

Güngör, N. D. (2003). *Brain drain from Turkey: An empirical investigation of the determinants of skilled migration and student non-return.* Unpublished PhD dissertations, Department of Economics, Middle East Technical University, Ankara, Turkey.

Güngör, N. D., & Tansel, A. (2008). Brain Drain from Turkey: An Investigation of Students' Return Intention. *Applied Economics*, 40, 3069–3087.

Güngör, N. D., & Tansel, A. (2014). Brain Drain From Turkey: Return Intentions of Skilled Migrants. *International Migration*, 52(5), 208–226.

CHAPTER 2

Brain Drain: Causes and Consequences

Abstract This chapter examines the causes and consequences of brain drain. The chapter reviews major migration theories and examines different perspectives on the brain drain and development. Finally, it reviews the literature on the gender dimension of brain drain.

Keywords Brain drain • Brain gain • Migration theories • Gender • Push-pull models

2.1 Introduction

This chapter aims to discuss the causes and consequences of the brain drain and its gender dimension. There exists a sizeable literature on the brain drain. The goal of the section is not to review such an immense literature by any means. There are several comprehensive literature surveys that focus on the different aspects of the brain drain (see inter alia Commander et al. 2002; Giannoccolo 2006; Oosterbeek and Webbink 2011; Katseli et al. 2006; Docquier and Rapoport 2012; Gibson and McKenzie 2012).

Before examining the literature on brain drain, it is worth taking a general look at the migration and development literature. Ravenstein (1885, 1889), the first scholarly works, considered migration basically as an economic phenomenon. This main perspective that people move from low-income to high-income regions is the basic notion of diverse approaches to migration.

2.1.1 The Neo-classical Migration Theory and New Economics of Labor Migration

According to the neo-classical economic theory, there are geographical differences in the supply and demand for labor, which causes wage differentials between regions. Therefore, individuals migrate from low-wage to high-wage areas. This is the macro-level explanation of migration decision. The theory sees migrants as all rational individual agents making a decision based on a cost-benefit calculation to earn the highest wages. This is the microlevel notion of the migration. The basis of the neo-classical migration theory is Harris-Todaro's two-sector model of rural-to-urban migration (Todaro 1969; Harris and Todaro 1970). This internal migration theory served as the basis for the main international migration theories including those of Todaro and Maruszko (1987) and Borjas (1989, 1990). The neo-classical migration theory suffers from a couple of fundamental issues. First, the theory is based on perfect factor markets, thereby fails to explain actual migration decisions of marginalized individuals with very limited or no access to factor markets in developing countries. Also, the theory does not deal with cultural, political, and institutional aspects of the migration decision. The theory has also been criticized for being ahistorical and Eurocentric in that it fails to acknowledge the different structural conditions of current developing countries and countries in the nineteenth- and twentieth-century Europe in terms of the framework of the transfer of labor from agricultural rural to industrial urban sectors (Skeldon 1997, cited in De Haas 2008).

The new economics of labor migration is an augmented version of neo-classical migration theory (Stark and Levhari 1982; Stark 1984; Stark and Bloom 1985; Massey et al. 1993). The new economics of labor migration theory challenges neo-classical models by considering family/household as the unit of analysis rather than individuals. People within families or households 'act collectively not only to maximize expected income but also to minimize risks and to loosen constraints associated with a variety of market failures, apart from those in the labor market' (Massey et al. 1993, p. 436). Another major difference of the new economics of labor migration models is that unlike the neo-classical models, remittances play a significant role in the decision to migrate.

2.1.2 The Dependency Theory

Historical-structural theory, adopting a Marxist political-economic approach, emerged as a response to neo-classical migration theory and developmentalist-modernizationist approaches toward development (Castles and Miller 2003; De Haas 2008). According to historical structuralists, individuals do not have a free choice in contrast to what neo-classical migration theory is based on. Individuals migrate not as a result of their free choice but rather they are forced to do so to move away from poverty, which is a result of the global political-economic structure.

The dependency theory, a part of the structuralist approach, underscores the interconnection of development and underdevelopment in the capitalist system. The theory argues that there are economic and political power imbalances among developed and underdeveloped countries, and this is an intrinsic part of the capitalist relations between these countries that reinforce inequalities among them (Frank 1966, 1969; Baran 1973). In this sense, migration, one manifestation of resource flow from peripheral to core countries, promotes the development of core countries at the expense of reinforced underdevelopment of poor countries in the periphery. These low-skilled immigrants from poor countries contribute to the prosperity of the core countries by becoming an urban proletariat, a source of cheap labor in the core countries. In the same manner, the migration of high-skilled workers is considered as a free transfer of human capital from underdeveloped countries, a waste of their investment in education.

De Haas (2008, p. 8) notes that '[h]istorical structuralists have been criticized for being too determinist and rigid in their thinking in viewing individuals as victims or "pawns" that passively adapt to macro-forces, largely ruling out individual agency'.

2.1.3 The Push-Pull Approach

The common shortcoming of neo-classical and dependency theory is that they cannot explain why some people in a certain country or region migrate while others do not (Massey et al. 1993; Reniers 1999). The push-pull model, the dominant migration model in the literature, is associated with Lee's (1966) framework that departs from Ravenstein's works. Lee suggests that the decision to migrate is determined by factors associated with the area of origin and destination and personal factors. In this sense, the push-pull model covers a wide range of factors that play a role in the decision to migrate. This basic characteristic is both its advantage and disadvantage. Although the model makes it possible to consider different factors, it fails to allow one to deal with these factors properly. Rather, the model covers different factors, individual and global, without assigning appropriate weights (De Haas 2008). Moreover, the model ignores the heterogeneity among individuals and stratification in societies in that a certain push or pull factor may have a different effect on the decision to migrate for different individuals. Similar to neo-classical models, the push-pull model has also been criticized for its narrow perspective, limiting decision to migrate to a cost-benefit analysis by individuals. The model ignores structural constraints that cause unequal access to resources (De Haas 2008).

Besides the new economics of labor migration, there are different approaches to explain return migration. For example, *the structural approach to return migration* contends that return decision should be analyzed regarding social and institutional factors in countries of origin, not individual experience per se. *Transnationalism* is more extended analysis that examines the important social and economic linkages between migrants' host and origin countries. Finally, *social network theory* underscores the role of social ties that migrants can create with other migrants.

2.2 Causes and Consequences

Brain drain is the movement of highly qualified individuals from their home countries to developed countries in order to reach greater opportunities in their field of specialization and/or to have better living condition and lifestyle. Migration of highly skilled workforce from developing countries to advanced countries is not a new phenomenon. Having reviewed four decades of economic research on the brain drain, Docquier and Rapoport (2012) conclude that 'high-skill migration is becoming a dominant pattern of international migration and a major aspect of globalization' (p. 724). The brain drain from developing countries has accelerated over the last decade due to the growth-of-information and knowledge-intensive activities. Also, the developed countries have actively encouraged the emigration of skilled labor by utilizing different incentives, including the use of temporary skilled migrant visas (Commander et al. 2002).

The term 'brain drain' first appeared in a report by the Royal Society of London in 1963 in the context of migration of British scientists to the US in the early 1960s. 'Brain' in 'brain drain' refers to any skill, competency, or attribute that is an asset. Drain 'implies that this rate of exit is at a greater level than "normal" or than what might be desired' (Giannoccolo 2006, p. 3). There are various terms that refer to 'brain drain' in the literature, such as 'highly qualified migration', 'skilled international migration', 'killed international labor circulation', 'professional transients', 'migration of expertise', 'quality migration', 'brain exodus', 'brain migration' or 'brain emigration', 'exodus of talent', and 'brain export'. Also, the flight of 'brain power' or 'loss of human capital' is used (Köser-Akçapar 2006).

Giannoccolo (2006) provides a comprehensive survey of the brain drain in the economic models. The first group of studies focuses on the effect of brain drain on the welfare of countries in terms of costs in the national accounts. The second group of studies deals with the role of education, and its effect on development, as education is the key element of innovation and technology, which are viewed as engines of economic growth. Another section of the literature examines the role of the brain drain in international commerce, considering its impact on some important production factors (p. 15). A significant portion of the literature focuses on the consequences of the brain drain on taxation.

The brain drain is considered to be one of the most detrimental aspects of international migration from the viewpoint of home country. The early studies in the brain drain literature in the 1960s argued that brain drain would reduce the overall welfare of the sending countries. This is because the high-skilled emigrants acquired partly public education. That is why Bhagwati and Hamada (1974) proposed a 'tax on brains' (Bhagwati tax) as emigrants' gain is the cost for those left behind (Oosterbeek and Webbink 2011).

By the 1960s and 1970s, the literature was mainly on the emigration of academics and professionals from developing countries. During this period, scholars of the brain drain started to use the term to also refer to the phenomenon of students preferring to stay in the developed countries where they had studied. Brandi notes that this phenomenon of 'unfair technological aid' from developing countries to the rich world was called 'reverse transfer of technology' in the United Nations Conference on Trade and Development in 1972 (Brandi 2001).

As the brain drain becomes the major desired pattern of international migration for the receiving countries, it turns out to be a prime concern for developing countries. Docquier and Rapoport (2012) state that globalization, via brain drain, leads to more divergence between countries in terms of human capital, creating higher income inequality between countries. Against brain drain, some demanded that developed countries should stop recruiting doctors from developing countries, and some developing countries restricted the migration of their highly skilled individuals (Gibson and McKenzie 2012).

There are three kinds of 'costs' of the brain drain for the sending country (Katseli et al. 2006). First, highly educated individuals lead to spill over benefits to other individuals. They lead to higher productivity in the economy. They are also likely to improve the governance and civic performance of the society. Second, a (perhaps large) part of education of emigrants is financed through fiscal revenues. Third, in the absence of those high-educated individuals there might be a difficulty in the delivery of health care and education. Substantial evidence showed that there exists a high correlation between the presence of highly educated persons and a well-functioning society, which is indicated by the lower incidence of poverty, political stability, a cleaner environment, greater income equality, lower crime rates, and lower population growth (McMahon 1999, cited in Katseli et al. 2006). Although the loss of scientists and engineers is important for developing countries, some scholars argue that 'given the growing importance of out-sourcing and the international diffusion of new technologies, at least some developing countries may gain more from scientists and engineers abroad than from those at home' (Eaton and Kortum 1996, 2002, cited in Katseli et al. 2006, p. 35).

Starting with the 1970s, some authors have investigated the linkage between brain drain and economic growth/development. Particularly, in the 1980s and 1990s studies focused on the long-run effects of the migration and of the education policies on economic growth. They incorporated human capital as an important productive factor. These issues were studied by both academia and some international organizations (Giannoccolo 2006).

Since the 1990s, a significant portion of scholars of the brain drain attempted to investigate the individual motivation to migrate by developing economic models with micro foundations. These models emphasized the role of wage differentials between home country and abroad. Some examined the 'optimal income taxation' in the context of migration. The literature in this period also dealt with the endogenous moving costs of the migration, the efficiency and equity issues in the relationship between HDCs and LDCs, the influence of the income on higher education, the debate about the private and public schools, the effect of income on the convergence in the industrial development of the countries, and urban growth (Giannoccolo 2006, p. 16). Commander et al. (2002) discuss the rationale of 'importing' skilled labor. Accordingly, this would lower wage costs and reduce domestic wage pressure. Also, importing skilled labor would widen the talent pool, easing the selection of the best candidates.

A paper by the United Nations Institute for Training and Research in 1978 dramatically changed the theoretical framework of the brain drain discussion as the paper stated that high-skilled emigrants returned home later (Brandi 2001). Since the late 1980s, scholars have shifted their attention from the migration of highly skilled individuals to 'labor market mobility'. So, seemingly the problem of brain drain in the short run becomes harmless *skill mobility* when viewed within a longer time perspective. In line with this shift in perspective, the terminology has changed to include 'brain circulation', 'brain gain', and 'brain exchange' (Köser-Akçapar 2006, p. 3). In the 1990s, the most common topic in the literature became the potential positive externalities for the sending countries, referred to as the 'brain gain'.

Docquier and Rapoport (2012) rightly argue that the brain drain creates winners and losers among developing countries. They go on to argue that 'certain source-country characteristics in terms of governance, technological distance, demographic size, and interactions between these, are associated with the ability of a country to capitalize on the incentives for human capital formation in a context of migration and to seize the global benefits from having a skilled, educated diaspora' (Docquier and Rapoport 2012, p. 725).

There are some specific gains for the sending countries (Commander et al. 2002; Katseli et al. 2006; Gibson and McKenzie 2012). First, the brain drain may raise human capital and thereby stimulate economic growth by providing a positive signal that motivates others in the sending country to acquire more education (Stark et al. 1997, 1998; Vidal 1998; Beine et al. 2001).

Second, the emigrants can contribute to the sending country through remittances or by providing essential inputs to new businesses and activities. At this point, it is important to note the case of Turkey, a major sending country with millions of citizens living abroad. Although channeling economic remittances to Turkey for economic development was a key strategy in the past, there is considerable decline in the importance of remittances as the economy grew significantly, and remittances turn out to be a negligible part of it. In this context, 'Foreign Exchange Deposit

Accounts with Credit Letter' and 'the Super Foreign Exchange Account' were introduced by the Central Bank of the Republic of Turkey in 1976 and 1994, respectively, in order to take advantage of the savings of the Turkish citizens living abroad (Çetin 2011). These accounts comprised of a significant portion of the Central Bank's foreign exchange reserves from 1976 to the early 2000s, easing the financial crises. Today, remittances are not as important as they used to be because Turkey has become well-integrated within the global economy, and the size of remittances is now negligible (Bettin et al. 2012; Bilgili and Siegel 2011). Currently remittances are only around 2 percent of the deficit, while they were almost half of the trade deficit in previous decades.[1]

Third, skilled emigration may increase the efficiency of the flow of knowledge and information. Major advances in communications technology reduce the extent to which skills are actually lost. That is, advanced technology creates collaboration between scholars across countries. However, Güngör (2003) rightly argues that although it is true that there have been improvements in individual productivity of academics resulting from such collaborations, what is equally (if not more) crucial is the return of scholars to fill the gap of (qualified) positions in the home country.

Gibson and McKenzie (2012), on the other hand, argued that high-skilled emigrants rarely engage in trade or foreign direct investment (FDI), but they do remit. In addition, the authors note, they rarely advise their governments or businesses in their home countries. Moreover, the evidence shows that while 'brain circulation' is more likely in the case of migration between developed countries, emigrants from developing countries are less likely to return to their origin countries (Cervantes and Guellec 2002). Beine et al. (2008) found that small states are the main losers of the brain drain because the brain drain involves a larger proportion of their skilled labor force and there are stronger reactions to the standard push factors. They conclude that small states are more successful in producing skilled individuals and less successful in retaining them. With Turkey, it was noted that highly skilled emigrants' knowledge exchange with their associates in Turkey is at low levels (Kaya 2002).

'Diaspora networks' is another way the brain drain can have a positive impact on the sending country. However, it is more likely that large countries can get the benefits from the diaspora networks. Lobbying is a crucial phenomenon and has been utilized intensively by some ethnic groups. For Turkey, on the other hand, it is safe to say that it is relatively a recent issue.

Overall, the empirical evidence provides inconclusive results on these possible effects. Docquier and Rapoport (2012) note that the controversy among economists over the sign and magnitude of the effects of brain drain on development is partly because of data constraints, which prevent these macro studies to provide evidence in a fully convincing way. Moreover, the authors note that although there is a bidirectional relationship between brain drain and development, there have only been studies that consider the unidirectional relationship.

2.3 Brain Drain and Gender Inequality[2]

A part of brain drain literature focuses on the causes of significant increase in the number of highly skilled women migrants in Organisation for Economic Co-operation and Development (OECD) countries from 5.7 million to 14.4 million between 1990 and 2010, surpassing the number of highly skilled men migrants (Pekkala et al. 2016, p. 5).

The increase in women's schooling rate is one key reason behind this pattern; the rise in education level increases the women's tendency to move (Dumont et al. 2007; Docquier et al. 2007a). This is confirmed by newer data provided by Docquier et al. (2009), suggesting that highly skilled women are emigrating at a rate higher than men. Their data shows that the average migration rates of females with postsecondary education are 17 percent higher than those of males. The authors contend that educated women are better able than uneducated women to escape gender discrimination in their home country.

Bang and Mitra (2011) support the role of access to education in brain drain by gender. They found that gender inequality in access to education and higher fertility rate, as a main reason for lower women's labor force participation rates and income levels, are both important push factors for women to flee their home country. That is, if women have equal access to education and lower fertility rates, they have less tendency to migrate. Bang and Mitra (2011) also showed that that the quality of political institutions has similar effects on brain drain of both genders. However, it is plausible to argue that such a finding may be problematic as a critical feminist perspective contends that 'neither gender differences in the experience of political, social and cultural conditions nor the career aspirations of highly skilled women are adequately taken into account' (Parvati 2009; Elveren and Toksöz 2017, p. 4).

Nejad (2013) looked at the effects of women's rights on the brain drain of highly skilled migration. She found contended even a small increase in a country's women's right index boosts the female brain drain rate. Nejad and Young (2014) confirmed this finding. Considering women's rights as a determinant of the female brain drain rate relative to that of men, they used women's economic, social, and political rights indices from Cingranelli and Richards (2010)'s Human Rights Dataset and migration flows across OECD and non-OECD countries to find that highly skilled women are more likely to migrate than men when women's rights levels are higher in the destination country than the origin country.

Some studies extend their analysis to cover all female migrants rather than just highly skilled ones to better understand the effect of the gap in women's rights between origin and destination countries on women's migration decisions. In this context, Ruyssen and Salomone (2015) used Gallup World Polls between 2009 and 2013 and measured gender discrimination in the proportion of female respondents stating that women in their country are not treated with respect and dignity and their desire to migrate, to conclude that '[o]verall we find that women who do not feel treated with respect and dignity in their country have a stronger desire to move out. Perceived gender discrimination hence positively affects the size of potential female migration' (Ruyssen and Salomone 2015, pp. 6–7, cited in Elveren and Toksöz 2017, p. 4). However, the important finding of this study is that highly skilled, employed, and secular women are more likely to turn their plans into action. In this regard, Iran is an important example. After Khomeini and the Islamic government took power in Iran, increasing number of educated people emigrated from Iran, and a large proportion of them were women escaping religious and ideological restrictions and gender-based discrimination (Chaichian 2011).

The gender aspect of the migration decision has been examined in other studies as well.[3] Nowak (2009) stated that worsening socio-economic conditions in Ghana caused flexibility in gender norms, which in turn increased migration among female nurses. Zweig and Changgui (1995), on scientists and engineers and students, showed a lower tendency of Chinese women residing in the US to return to China as they have more opportunities for career advancement in the US (cited in Murakami 2009). Ono and Piper (2004) and Murakami (2009) note a similar situation with Japan.

Güngör (2014), based on dataset of Tansel and Güngör (2002) and Güngör (2003), examines the gender dimension of return intentions of skilled labor in Turkey. She shows that women are more likely to migrate due to greater opportunities for advanced training and work and because they have a concern about advancing in their jobs in the home country. Güngör (2014) also underscores that women are more likely to migrate and not return because they are more affected by economic decline and lack of social security in Turkey. 'Political discord, bureaucracy and lack of financial resources to conduct business in Turkey, on the other hand', she shows that 'appear to be significantly more important push factors for male respondents' (Güngör 2014, p. 118). She also notes that women are more dependent on men in migration decision. This short article concludes that 'female professionals indeed have a higher, statistically significant probability of not returning than male professionals, controlling for relevant factors' (ibid., p. 118).

Spadavecchia (2013) argues that women from sub-Saharan Africa move to Europe to escape from strict gender roles that confine them to the household and restrict their access to credit, land, and means of production. A very similar case for professional women in Nigeria is noted by Reynolds (2006). Using a data set prepared by Statistics Lithuania covering Lithuania's entire population in 2011 and linked with data on emigration, Klüsener et al. (2015) found that persistent gender inequality is one possible determinant of higher tendency of educated women to emigrate than their male peers. De Jong (2000) shows that gender norms of caregiving have an impact on migration decisions for men and women in Thailand. While having dependent children and elderly adults in the household reduce women's migration intentions, they increase men's intention to migrate. Rangelova (2006) showed that women are more likely to resettle abroad than men. Finally, Alberts and Hazen (2005) note that differing gender roles between the US and the home countries of international students have some impact on their (non-)return intentions.

Overall, the above literature shows that gender inequality is a significant push factor for women. However, another part of the brain drain literature with a focus on the role of gender inequality does not confirm this general argument. For example, based on original indices for gender inequality in labor market and migration data provided by Docquier et al. (2009) covering 1991–2001, Baudassé and Bazillier (2014) found that gender inequality in origin countries is not a significant push factor for women. Similarly, based on data obtained from Botswanan students in their final year in tertiary education, Campbell (2007) found that gender differences do not have a significant effect on intention to emigrate. Finally, Bartolini et al. (2017), based on a survey conducted in 2013, also found no general gender differences in migration decisions.

Notes

1. See Özden (2005) and Özden and Schiff (2005) for the discussion on the remittances in international migration and Bettin et al. (2012) for a comprehensive work on the remittances in the case of Turkey.
2. This section mainly comes from Elveren and Toksöz (2017).
3. McKenzie et al. (2014) on Filipino migrants found that women are slightly more responsive to gross domestic product (GDP) shocks at destination.

References

Alberts, H. C., & Hazen, H. D. (2005). "There are always two voices…": International Students' Intentions to Stay in the United States or Return to their Home Countries. *International Migration, 43*(3), 131–154.

Bang, J. T., & Mitra, A. (2011). Gender bias and the female brain drain. *Applied Economics Letters, 18*, 829–833.

Baran, P. (1973). On the political economy of backwardness. In C. K. Wilber (Ed.), *The Political Economy of Development and Underdevelopment* (pp. 82–93). New York: Random House.

Bartolini, L., Gropas, R., & Triandafyllidou, A. (2017). Drivers of highly skilled mobility from Southern Europe: Escaping the crisis and emancipating oneself. *Journal of Ethnic and Migration Studies.* https://doi.org/10.1080/1369183X.2016.1249048

Baudassé, T., & Bazillier, R. (2014). Gender inequality and emigration: Push factor or selection process? *International Economics, 139*, 19–47.

Beine, M., Docquier, F., & Rapoport, H. (2001). Brain Drain and Economic Growth: Theory and Evidence. *Journal of Development Economics, 64*, 275–289.

Beine, M., Docquier, F., & Schiff, M. (2008). *Brain Drain and its Determinants: A Major Issue for Small States.* IZA Discussion Paper No. 3398.

Bettin, G., Elitok, S. P., & Straubhaar, T. (2012). Causes and Consequences of the Downturn in Financial Remittances to Turkey: A Descriptive Approach. In S. P. Elitok & T. Straubhaar (Eds.), *Turkey, Migration and the EU: Potentials, Challenges and Opportunities.* Hamburg: Hamburg University Press.

Bhagwati, J., & Hamada, K. (1974). The Brain Drain, International Integration of Markets for Professionals and Unemployment: A Theoretical Analysis. *Journal of Development Economics, 1*, 19–42.

Bilgili, Ö., & Siegel, M. (2011). *Understanding the changing role of the Turkish diaspora*. UNU-MERIT Working Paper Series 2011-039.

Borjas, G. J. (1989). Economic Theory and International Migration. *International Migration Review*, 23, 457–485.

Borjas, G. J. (1990). *Friend or Strangers: The Impact of Immigrants on the US Economy*. New York: Basic Books.

Brandi, M. C. (2001). *The evolution in theories of the brain drain and the migration of skilled personnel*. Institute for Research on Population and Social Policies—National Research Council, Rome, Italy.

Campbell, E. K. (2007). Brain Drain Potential in Botswana. *International Migration*, 45(5), 115–145.

Castles, S., & Miller, M. J. (2003). *The Age of Migration*. Hampshire and London: Macmillan.

Cervantes, M., & Guellec, D. (2002). The Brain Drain: Old Myths, New Realities. *OECD Observer*. http://oecdobserver.org/news/archivestory.php/aid/673/The_brain_drain:_Old_myths,_new_realities.html. Accessed 14 March 2018.

Çetin, B. (2011). 35. Yılında Merkez Bankası Kredi Mektuplu Döviz Tevdiat Hesabı Sistemi. *İstihdamda 3 İ İşgücü, İşveren, İşkur*, 62–63.

Chaichian, M. A. (2011). The new phase of globalization and brain drain: Migration of educated and skilled Iranians to the United States. *International Journal of Social Economics*, 39(1/2), 18–38.

Cingranelli, D. L., & Richards, D. L. (2010). *The Cingranelli-Richards (CIRI) Human Rights Dataset Version 2010.05.17*.

Commander, S., Kangasniemi, M., & Winters, L. A. (2002). *The Brain Drain: Curse or Boon? A Survey of the Literature*. Paper prepared for the CEPR/NBER/SNS International Seminar on International Trade, Stockholm, May 24–25, 2002.

De Haas, H. (2008). *Migration and development: A theoretical perspective*. International Migration Institute Paper No. 9, University of Oxford.

De Jong, G. F. (2000). Expectations, gender, and norms in migration decision-making. *Population Studies*, 54(3), 307–319.

Docquier, F., Lohest, O., & Marfouk, A. (2007a). Brain Drain in Developing Countries. *World Bank Economic Review*, 21(2), 193–218.

Docquier, F., Lowell, B. L., & Marfouk, A. (2009). A Gendered Assessment of Highly Skilled Emigration. *Population and Development Review*, 35(2), 297–321.

Docquier, F., & Rapoport, H. (2012). Globalization, Brain Drain, and Development. *Journal of Economic Literature*, 50(3), 681–730.

Dumont, J.-C., Martin, J. P., & Spielvogel, G. (2007). *Women on the Move: The Neglected Gender Dimension of the Brain Drain*. IZA Discussion Paper No. 2920, IZA, Bonn, Germany.

Eaton, J., & Kortum, S. (1996). Measuring Technology Diffusion and the International Sources of Growth. *Eastern Economic Journal*, 22(4), 401–410.

Eaton, J., & Kortum, S. (2002). Technology, Geography, and Trade. *Econometrica*, 70(5), 741–779.
Elveren, A. Y., & Toksöz, G. (2017). Why Don't Highly Skilled Women Want to Return? Turkey's Brain Drain from a Gender Perspective. MPRA No. 80290, 2017. https://mpra.ub.uni-muenchen.de/80290/
Frank, A. G. (1966). *The Development of Underdevelopment*. New York: Monthly Review Press.
Frank, A. G. (1969). *Capitalism and Underdevelopment in Latin America*. New York: Monthly Review Press.
Giannoccolo, P. (2006). *The Brain Drain: A Survey of the Literature*. Università degli Studi di Milano-Bicocca, Department of Statistics, Working Paper No. 2006-03-02. http://ssrn.com/abstract=1374329 or https://doi.org/10.2139/ssrn.1374329
Gibson, J., & McKenzie, D. (2012). The economic consequences of 'Brain Drain' of the best and brightest: Microeconomic evidence from five countries. *The Economic Journal*, 122(560), 339–375.
Güngör, N. D. (2003). *Brain drain from Turkey: An empirical investigation of the determinants of skilled migration and student non-return*. Unpublished PhD dissertation, Department of Economics, Middle East Technical University, Ankara, Turkey.
Güngör, N. D. (2014). *The Gender Dimension of Skilled Migration and Return Intentions*. Invited presentation, Harnessing knowledge on the migration of highly skilled women: An expert group meeting, International Organization for Migration and OECD Development Center, Geneva, Switzerland, April 3–4, 2014.
Harris, J. R., & Todaro, M. (1970). Migration, unemployment and development: A two-sector analysis. *American Economic Review*, 60, 126–142.
Katseli, L. T., Lucas, R. E. B., & Xenogiani, T. (2006). *Effects of Migration on Sending Countries: What Do We Know?* OECD Development Centre, WP No. 250.
Kaya, M. (2002). *Beyin göçü/erozyonu*. Technology Research Centre Report, Osmangazi University, Eskişehir, Turkey.
Klüsener, S., Stankunienee, V., Grigoriev, P., & Jasilionis, D. (2015). The Mass Emigration Context of Lithuania: Patterns and Policy Options. *International Migration*, 53(5), 179–193.
Köser-Akçapar, Ş. (2006). Do Brains really going down the Drain? Highly skilled Turkish Migrants in the USA and the «Brain Drain» Debate in Turkey. *Revue européenne des migrations internationales*, 22(3), 79–107.
Lee, E. S. (1966). A Theory of Migration. *Demography*, 3, 47–57.
Massey, D. S., Arango, J., Hugo, G., Kouaouci, A., Pellegrino, A., & Taylor, J. E. (1993). Theories of international migration: A review and appraisal. *Population and Development Review*, 19, 431–466.
McKenzie, D., Theoharides, C., & Yang, D. (2014). Distortions in the International Migrant Labor Market: Evidence from Filipino Migration and Wage Responses

to Destination Country Economic Shocks. *American Economic Journal: Applied Economics*, 6(2), 49–75.
McMahon, W. M. (1999). *Education and Development: Measuring the Social Benefits*. New York: Oxford University Press.
Murakami, Y. (2009). Incentives for International Migration of Scientists and Engineers to Japan. *International Migration*, 47(4), 67–91.
Nejad, M. N. (2013). *Institutionalized Inequality and Brain Drain: An Empirical Study of the Effects of Women's Rights on the Gender Gap in High-Skilled Migration*. IZA Discussion Paper No. 7864.
Nejad, M. N., & Young, A. T. (2014). *Female Brain Drains and Women's Rights Gaps: A Gravity Model Analysis of Bilateral Migration Flows*. IZA Discussion Paper No. 8067.
Nowak, J. (2009). Gendered perceptions of migration among skilled female Ghanaian nurses. *Gender & Development*, 17(2), 269–280.
Ono, H., & Piper, N. (2004). Japanese Women Studying Abroad: The Case of the United States. *Women's Studies International Forum*, 27, 101–118.
Oosterbeek, H., & Webbink, D. (2011). Does Studying Abroad Induce a Brain Drain? *Economica*, 78, 347–366.
Özden, Ç. (2005). Educated migrants: Is there brain waste? In Ç. Özden & M. Schiff (Eds.), *International Migration, Remittances and the Brain Drain* (pp. 227–224). Washington, DC: The World Bank and Palgrave Macmillan.
Özden, Ç., & Schiff, M. (Eds.) (2005). *International Migration, Remittances and the Brain Drain*. Washington, DC: The World Bank and Palgrave Macmillan.
Parvati, R. (2009). Situating women in the brain drain discourse: Discursive challenges and opportunities. In H. Stalford, S. Currie & S. Velluti (Eds.), *Gender and Migration in 21st Century Europe* (pp. 85–106). Aldershot: Ashgate.
Pekkala, K. S., Kerr, W., Özden, Ç., & Parsons, C. (2016). *Global Talent Flows*. NBER Working Paper Series 22715.
Rangelova, R. (2006). Gender Dimension of the New Bulgaria's Migration: Comments on Empirical Data. http://aa.ecn.cz/img_upload/9e9f2072be82f3d69e3265f41fe9f28e/RRangelova_Gender_Dimension.pdf
Ravenstein, E. G. (1885). The Laws of Migration. *Journal of the Royal Statistical Society*, 48, 167–227.
Ravenstein, E. G. (1889). The Laws of Migration. *Journal of the Royal Statistical Society*, 52, 214–301.
Reniers, G. (1999). On the History and Selectivity of Turkish and Moroccan Migration to Belgium. *International Migration*, 37, 679–713.
Reynolds, R. R. (2006). Professional Nigerian Women, Household Economy, and Immigration Decisions. *International Migration*, 44(5), 167–188.
Ruyssen, I., & Salomone, S. (2015). Female migration: A way out of discrimination? https://www.cesifo-group.de/dms/ifodoc/docs/Akad_Conf/CFP_CONF/CFP_CONF_2015/cemir15-Poutvaara/Papers/cemir15_Salomone.pdf
Skeldon, R. (1997). *Migration and Development: A Global Perspective*. Essex: Longman.

Spadavecchia, C. (2013). Migration of Women from Sub-Saharan Africa to Europe: The Role of Highly Skilled Women. *Sociología y tecnociencia/Sociology and Technoscience*, 3(3), 96–116.

Stark, O. (1984). Migration decision making: A review article. *Journal of Development Economics*, 14, 251–259.

Stark, O., & Bloom, D. E. (1985). The New Economics of Labor Migration. *American Economic Review*, 75, 173–178.

Stark, O., & Levhari, D. (1982). On migration and risk in LDCs. *Economic Development and Cultural Change*, 31, 191–196.

Stark, O., Helmenstein, C., & Prskawetz, A. (1997). A Brain Gain with a Brain Drain. *Economics Letters*, 55, 227–234.

Stark, O., Helmenstein, C., & Prskawetz, A. (1998). Human Capital Depletion, Human Capital Formation, and Migration: A Blessing or a Cure? *Economics Letters*, 60, 363–367.

Tansel, A., & Güngör, N. D. (2002). 'Brain Drain' from Turkey: Survey Evidence of Student Non-Return. http://ssrn.com/abstract=441160

Todaro, M. P. (1969). A model of labor migration and urban unemployment in less-developed countries. *American Economic Review*, 59, 138–148.

Todaro, M. P., & Maruszko, L. (1987). Illegal migration and US immigration reform: A conceptual framework. *Population and Development Review*, 13, 101–114.

Vidal, J.-P. (1998). The Effect of Emigration on Human Capital Formation. *Journal of Population Economics*, 11, 589–600.

Zweig, D., & Changgui, C. (1995). *China's Brain Drain to the United States: Views of Overseas Chinese Students and Scholars in the 1990s*. China Research Monograph, Institute of East Asian Studies, University of California, Berkeley, CA.

CHAPTER 3

Brain Drain in Turkey: A Literature Review

Abstract This chapter provides a comprehensive literature review on three parts of brain drain in Turkey. First, it covers nonempirical works and development plans; second, it reviews empirical works; and finally, it deals with current debates, focusing on reverse brain drain. The chapter discusses the pull and push factors to draw a detailed picture of brain drain from Turkey, from the 1960s to today.

Keywords Brain drain • Reverse brain drain • High-skilled migration • Higher education • Economic development

The brain drain is the departure of high-skilled professionals to work (mostly) in advanced countries. It was in the 1960s that high-level professionals (i.e., doctors and engineers) migrated mostly to the Western Europe to work. They were the early examples of the 'brain drain' phenomenon in Turkey.

Turkey has long been one of the top ten countries that send their students to pursue graduate degrees in the US, along with much more populous India and China. According to 'the Doctoral Recipients from US Universities' report, Turkey is sixth in a list of foreign citizens who earned a doctoral degree from a US institution between 2004 and 2014. When India and China are excluded (or in terms of per capita), Turkey is fourth in the list.[1]

It is important for the policymakers in Turkey to focus on this issue as Turkey is a fast developing country with a great potential in terms of human capital. That is, Turkey can obtain the maximum benefit from these emigrants, either by encouraging them to return to the country or by creating efficient linkages with them in their host countries. Social scientists in Turkey have paid significant attention to the migration of high-skilled labor to understand the causes and consequences of the so-called brain drain. The purpose of this chapter is to examine the works on the brain drain in detail. To this end, the following sections aim at providing a literature review on the brain drain in Turkey. Section 3.1 deals with nonempirical literature in the brain drain and discusses the 'Five-Year Development Plans' in terms of their approach to the brain drain. Section 3.2 focuses on the empirical works, and Sect. 3.3 examines the current debates and the phenomenon of so-called reverse brain drain.

3.1 Nonempirical Literature on the Brain Drain

Turkey has over 4.5 million citizens living abroad. Along with Morocco and South European countries, Turkey was a major emigration country for Western European countries in the 1960s and 1970s, particularly providing low-skilled labor for Germany and the Netherlands. There are currently more than three million Turkish nationals residing in the EU, over two million of which in Germany. The US, France, the Netherlands, and Austria are other major countries with high population of Turkish nationals.[2] Among these countries it is the US—along with Canada—in which the most skilled migrants are located. While only a quarter of the highly skilled individuals migrate to Europe, two-thirds prefer North America, particularly the US (Katseli et al. 2006). Although Turkey is the leading country (with a share of 5.8 percent) who have sent migrants to EU, followed by Morocco (4.5 percent), Algeria (3.9 percent), Serbia and Montenegro (2.2 percent), India (1.8 percent), Albania (1.7 percent), Tunisia (1.3 percent), and Pakistan (1.2 percent, respectively), Turkey is not a major country in terms of high-skilled emigrants. The main high-skilled labor-sending countries are Algeria (13.5 percent), Morocco (3.1 percent), and India (2.7 percent). Turkey's share is as low as 1.4 percent (Katseli et al. 2006).

There is a sizable literature on migration in Turkey (see inter alia Toksöz 2006; Dedeoğlu and Gökmen 2011; Paçacı Elitok and Straubhaar 2012).[3] A significant part of this literature focuses on the migration toward

European countries (Toksöz 2006; Dedeoğlu 2014), particularly to Germany.[4] The scholars started to pay attention to the migration in the late 1960s when the first-generation immigrants began to fill jobs in Germany and other major European countries (Abadan Unat 2002; Toksöz 2006). Yiğit (2011) reports the general pattern of this migration. Accordingly, about 800,000 individuals were sent by İŞKUR (the Employment Agency, İş ve İşçi Bulma Kurumu, the name by then) to Germany, Australia, Austria, Belgium, France, the Netherlands, and Switzerland (about 650,000 of them were sent to Germany between 1961 and 1975; the father of the author of this book was one of them). There was increase in migration to oil-rich Arab countries particularly to Libya and Saudi Arabia, beginning with the second half of 1970s as a result of the economic crises and following recruitment stop in European countries in the first half of 1970s. Those emigrants were mostly recruited by the construction companies. This pattern slowed down due to the Gulf War that begun in 1991. İŞKUR send about 700,000 workers to Middle East and North African countries between 1975 and 1994 (Yiğit 2011). The major host countries in this period were Saudi Arabia, Libya, Iraq, Kuwait, Yemen, and Jordan. The final major destination of migration became Russia and Central Asian Turkic countries in the 1990s.

3.1.1 Institutional Structure and Development Plans

Considering the increasing population of the Turkish citizens residing abroad, Turkey has expanded her relations with its emigrants (Bilgili 2012; Bilgili and Siegel 2011, 2013). Tanıtma Fonu Kurulu Başbakanlık Merkez Teşkilatı (The Promotion Fund of the Office of the Turkish Prime Minister) was established in 1985. Later in 1998 Yurtdışı Vatandaşlar Danışma Kurulu (Consultancy Board for Citizens Living Abroad) was established. Dış İlişkiler ve Yurtdışı İşçi Hizmetleri Genel Müdürlüğü (Foreign Relations and Workers Abroad Services General Directorate) has expanded, and Yurt-Danış Bürosu (Homeland-Advice Bureau) was founded in 2001. Moreover, in order to increase the efficiency of assistance to the Turkish citizens living abroad, Yurtdışı Türkler ve Akraba Topluluklar Başkanlığı (Presidency for Turks Abroad and Related Communities) was established with a new law that was approved in March 2010. The organization performs under the Office of the Prime Minister and in coordination with other ministries and governmental organizations that are involved with Turks abroad (Bilgili 2012; Yurtdışı Türkler ve Akraba Topluluklar Başkanlığı 2018).

The approach of policymakers to the phenomenon of brain drain can be best tracked by reviewing the Five-Year Development Plans. Atılgan (1986) and Tuncel (2003) examine 'the Five-Year Development Plans' in detail. Accordingly, the First Five-Year Development Plan (1963–1967) considers the migration in terms of economic development, remittances, and the prevention of unemployment (DPT 1963). The plan views the export of excess labor force (i.e., low-skilled workers) as a positive issue in terms of economic development.

The Second Five-Year Development Plan (1968–1972), however, notes that 38 percent of immigrants were 'skilled labor' (DPT 1967). Therefore, the plan stresses that there is a need to prevent the migration of skilled labor. Atılgan (1986) and Tuncel (2003) note that it is the Third Five-Year Development Plan (1973–1977) that includes the term 'brain drain' for the first time. The plan states that 'by 1970 7 per cent of architects, 5.3 per cent of mechanical engineers, 8.2 per cent of specialists, and 21.4 per cent of general practitioners moved abroad to work' (DPT 1972, p. 81). In other words, the plan seems to be concerned about the brain drain.

The Fourth Five-Year Development Plan (1979–1983) is the last plan that notes the negative impact of brain drain on Turkey as a developing country (DPT 1979, 1981; Atılgan 1986; Tuncel 2003). The Fifth Five-Year Development Plan (1985–1989) only notes the shortage of medical stuff (except for pharmacists), electrical engineers, computer engineers, teachers, and academics, without referring to the concept of brain drain or any clear concern about it (DPT 1985a, b).

The Sixth Five-Year Development Plan (1990–1994) and the Seventh Five-Year Development Plan (1996–2000) do not mention brain drain, just noting the shortage of some skilled labor workforce and the necessity of providing required training in order to develop human capital (DPT 1989, 1995).

The Eighth Five-Year Development Plan (2001–2005) once again refers to the concept of brain drain, pointing to the need for introducing measures to prevent the brain drain (DPT 2000a, b, pp. 113–115). While the Ninth Plan (2007–2013) does not mention brain drain (DPT 2006), the Tenth Plan (2014–2018) suggests that brain migration toward Turkey in desired fields should be encouraged particularly from the countries in our region. The plan stresses that this will improve the quality of human capital, which will contribute to Turkey's development capacity (DPT

2013, p. 10). However, the plan notes that compared to advanced countries, Turkey does not take full advantage of 'the high skilled brain migration' (ibid., p. 188).

3.1.2 Nonempirical Literature on the Brain Drain

Having examined the approach of the Five-Year Development Plans to the brain drain, we can now return our attention to the general literature on the brain drain in Turkey. There are some studies that discuss the term 'brain drain', providing some policy recommendations for Turkey.[5] The main reason for the brain drain in Turkey is that the economy cannot create enough jobs for the young population in Turkey. Although highly skilled migration has always been a well-known phenomenon in Turkey, policymakers began to pay more attention to the issue with the economic crisis that Turkey experienced at the beginning of the millennium (Köser-Akçapar 2009). The impact of such a severe crisis was very remarkable. An early study on this issue draws a clear picture in this context. According to a survey conducted by the Society for Human Resource Management (Cumhuriyet Gazetesi 07.05.2002), 7 percent of high-skilled individuals who lost their jobs during the 2001 economic crises went abroad. The same study also revealed that 12 percent of the same group were trying to emigrate, 8 percent went abroad to pursue a graduate degree (i.e., master's) or attend foreign language program while they are unemployed, and finally, 30 percent of them noted that they were willing to go abroad once they had a chance (Işığıçok 2002).

Tuncel (2003) in a comprehensive discussion, in addition to commonly cited types of migration, remarks also on so-called political migration, mostly associated with the military coup of 1980. This type of migration has received virtually no attention from scholars in Turkey. There are different estimations on the number of political immigrants of this period, ranging between 14,000 and 30,000 (Çizmeci 1988, p. 35, cited in Tuncel 2003). As a matter of fact, it is argued that most of these immigrants were highly skilled individuals; so it is safe to consider them to be a part of the brain drain (Abakay 1988, pp. 1–16, cited in Tuncel 2003).

Most of these aforementioned studies have a similar general discussion on the brain drain phenomenon and provide similar policy recommendations. Here, we would like to highlight some points emphasized by various authors. Gündoğdu (2009), for instance, focused on the direct linkage between preventing brain drain and economic development. Similarly,

Barışık and Çetinbaş (2009) examined the linkage between the level of research and development expenditures and the brain drain in detail. Most of these studies deal with how to prevent the brain drain. Kaya (2002, 2009), for instance, provides a detailed discussion on the brain drain in Turkey from a historical perspective and suggests policy recommendations to prevent it. Cansız (2006) proposes some general measures to prevent or compensate the brain drain loss. He suggests that the loss of brain drain may be compensated for by encouraging qualified students from the Turkic countries to pursue their higher education in Turkey. He also points out the time gap between establishing new departments/universities and recruiting the qualified faculty members for available positions in these new departments. He suggests encouraging foreign instructors to serve for these positions until they can be replaced by those who were sent abroad by the government. Finally, the author emphasizes the autonomy/independency of the universities in terms of efficiency of the use of the resources they have. Similarly, Özdemir (2009) discusses the issue of brain drain at a global scale and emphasizes some widely cited reasons for brain drain in Turkey, namely, problems with higher education system. Çengel (2009) also notes some issues in the context of brain drain and provides some suggestions. He argues for the importance of the freedom of conscience and religion, freedom of speech and academic freedom, and extending the boundaries of individual freedom. The author suggests putting an end to the compulsory military service and permitting foreign doctors to be able to work in Turkey. Fişek (2009), on the other hand, argues that the structure of the higher education system serves the brain drain, because it was designed by global actors. Güngör (2010) asserts that although it is argued that the branches of American universities across the world may reduce brain drain and increase the quality level, this also comes with its cost, as it may erode the national cultural identity.

In a recent study, Alvan (2012) states that one key reason for brain drain is problems in the higher education system. The author argues that there are some shortcomings in the higher education system in Turkey, such as a mismatch between available jobs, the skills required in the market, and the courses available in the higher education system. That is, students are not well-trained for the labor market. She underscores the wage differential between the home country and abroad. She argues that there is a mismatch between the number of skilled persons demanded by the public and private sectors and the number of university graduates. The author suggests substantial structural renewal of the education system and improvements in the recruitment system in order to reduce brain drain.

Another part of this literature pays attention to the push and pull factors of the brain drain. For instance, Sağbaş (2009) in her study reveals some not unexpected facts about the socio-economic background of highly skilled migrants. Accordingly, those highly skilled professionals are young, single, male, and from a higher social strata. Two main reasons to move abroad are 'professional issues' and 'desire to reach higher income'. Sağbaş' study reveals that the primary factors in return decisions are 'improved occupational conditions' and 'economic and political stability' (Sağbaş 2009, cited in Vatansever Deviren and Daşkıran 2014).

Similarly, Babataş (2007) examines the pull and push factors of the brain drain. The study reveals some common factors discussed earlier in the literature. Accordingly, push factors include unemployment, lower wages, economic instability and concern for the future, political instability, lack of freedom of speech and academic freedom, problems with bureaucracy, inequality in educational opportunities, unplanned higher education, and advantage of having education in a foreign language in adapting to abroad. Pull factors, on the other hand, include better life standards, higher income, better research environment, better life conditions, greater opportunities for better education, higher technological level, need for skilled labor, and better income distribution (Babataş 2007, cited in Vatansever Deviren and Daşkıran 2014).

The goal of this section was to review the general literature on the brain drain in Turkey in 2000s. The next section deals with the empirical works on the brain drain.

3.2 Empirical Works

Although there has been sizeable literature on migration issue for Turkey, the volume of works that focuses on brain drain per se is not significant. Especially, there are a few comprehensive empirical works that investigate the migration of highly skilled work power. Below is an attempt to review this literature (see Table 3.1).

Most of these studies focus on the US, the top destination country for the high-skilled individuals. Köser-Akçapar (2006) notes that there are different statistics on the number of Turkish students in the US (p. 7). Accordingly, while the 2005 Institute of International Education Report

Table 3.1 Empirical literature on the brain drain in Turkey

Study	Period of study	Target group/sample	Number of participant	Source of data/method
Dirican et al. (1968)	1968	Medical doctors working abroad	230	Questionnaire
Kösemen (1968)	1968	Engineers and architects working abroad	975	Questionnaire
Oğuzkan (1971)	1968–1969	Scholars with doctoral degrees working abroad	150	Questionnaire
The Turkish Chambers of Engineers and Architects (1972)	1972	Engineers and architects working in Turkey	7120	Questionnaire
Uysal (1972)	1972 (see Target group/sample)	Recipients of government scholarships from 1929 to 1972, working in Turkey or abroad	3232	Questionnaire and data from archives
Kurtuluş (1999)	1991	Students in the US	90	Questionnaire
Yavuzer (2000)	1998–2000	General Turkish population living in Washington D.C., Maryland, and Virginia	363	Questionnaire and interviews
Öztürk (2001)	1997–1998	Students in the US	330	Questionnaire, in-depth interviews, regression analysis, and one-tailed t-tests
Sunata (2002)	2002	The information and communication technology experts	178	Questionnaire, in-depth interviews
Altaş et al. (2006)	2005–2006	Academicians (with a degree from a foreign institution) from 27 universities in Turkey	200	Questionnaire, a loglinear model
Köser-Akçapar (2006)	2005–2006	Graduate students and professionals in the US, returnees	140	All available secondary data, semi-structured and in-depth interviews

(*continued*)

Table 3.1 (continued)

Study	Period of study	Target group/sample	Number of participant	Source of data/method
Güngör and Tansel (2008a)	2002	Undergraduate and graduate students residing abroad	1103	Questionnaire, ordered probit model
Güngör and Tansel (2008b)	2002	Professionals residing abroad	1224	Questionnaire, correspondence analysis
Gökbayrak (2009)	2004–2005	Permanent resident engineers abroad	130	Questionnaire, Chi-square tests and correlation analysis, in-depth interviews
Pazarcık (2010)	Dec. 2009–Feb. 2010	Academicians from top US universities	50	Questionnaire, Chi-square tests, and correlation analysis
Akman (2014)	Dec. 12–13, 2009	Undergraduate students (mostly) from Kocaeli University	210	Questionnaire, the analysis of variance
Esen (2014)	Jun. 2013–Oct. 2013	The US-educated scholars teaching in the US or in Turkey	20	In-depth interviews
Güngör (2014)	2002	Students and professionals residing abroad	2327	Questionnaire, ordered probit model
Güngör and Tansel (2014)	2002	Professionals residing abroad	1224	Questionnaire, ordered probit model
Mollahaliloğlu et al. (2014)	Sep. 2009–Nov. 2009	Physicians applying for a position at the Ministry of Health	3690	Questionnaire, logistic regression analysis
Vatansever Deviren and Daşkıran (2014)	2014	Faculty of Muğla Sıtkı Koçman University (holds a degree from or has been in a foreign university)	106	Questionnaire
Elveren and Toksöz (2017)	Oct. 2015–Feb. 2016	Undergraduate and graduate students and professionals residing abroad	200	Questionnaire, Chi-square tests, correlation analysis, ordered probit model

Source: Author's compilation

states that there are 12,474 Turkish students in the US, the same number is 15,000 according to Louscher and Cook's (2001) estimations, and it is 14,518 according to the 2004 Yearbook of Immigration Statistics released in January 2006 by the US Homeland Security, while the Student and Exchange Visitor Program of the US Immigration and Customs Enforcement indicates that there are a total of 13,923 Turkish students in the US. Also, while there are about 22,000 self-funded students or supported by their parents as reported by the Turkish Ministry of National Education, this number is nearly double according to the United Nations Educational, Scientific and Cultural Organization (UNESCO) (Güngör 2010).

The empirical works on the brain drain in Turkey listed in Table 3.1 and reviewed in detail below can be categorized in four groups. The first group includes the Turkish Chambers of Engineers and Architects (1972), Akman (2014), and Mollahaliloğlu et al. (2014), which target those who are in Turkey, mostly with no degree received from a foreign institution. Therefore, although they consider numerous persons (7120 in the Turkish Chambers of Engineers and Architects [1972] and 3690 in Mollahaliloğlu et al. [2014]) or present detailed statistical analysis as in the case of Mollahaliloğlu et al. (2014), their findings provide limited information about the phenomenon of the brain drain in Turkey.

The second group in the literature includes Altaş et al. (2006) and Vatansever Deviren and Daşkıran (2014), which targets returnees. Although they consider only academicians, ignoring those in the private sector, Altaş et al. (2006) and Vatansever Deviren and Daşkıran (2014) provide some valuable insights into the Turkish brain drain.

The third group includes Uysal (1972), Köser-Akçapar (2006), and Esen (2014), who target both stayers and returnees. Even though a small group of people (20 in Esen 2014) are included, these studies provide substantial information since some of them utilize the in-depth interview method to compare the decisions of stayers and returnees.

The fourth group is the studies that focus on students and/or professionals abroad. In addition to Yavuzer (2000) and Pazarcık (2010), this group includes the early examples of empirical works, namely, Dirican et al. (1968), Kösemen (1968), Oğuzkan (1971), and Kurtuluş (1999), some major works such as Öztürk (2001) and Gökbayrak (2009), and the most important contributions to the literature, namely, Güngör and Tansel (2008a, b, 2014), which are reviewed in Sect. 3.2.1.

- *The first group: targets those who are in Turkey, mostly with no degree received from a foreign institution*

The Turkish Chambers of Engineers and Architects (1972) is reviewed by Başaran (1972a, b). The study is based on a questionnaire filled by 7120 engineers and architects, conducted by the Turkish Chambers of Engineers and Architects in 1972 in Ankara. The study revealed that there is a very high tendency to migrate among engineers and architects if they had a job offer, as high as 71.5 percent among the public sector employees, 58.7 percent among the private sector employees, and 44.8 percent among employers in private sector. It is important to note that the ratio of tendency to migrate even if there was no job offer was also higher than 30 percent. The main reason for migration was lower wages at home (or possibility of having higher wages abroad).

Akman (2014) examines the students' perceptions or intentions to migrate with a survey conducted face-to-face, employing the analysis of variance (ANOVA). The study is based on a survey of 210 out of 350 students (virtually all of whom are students at Kocaeli University) who attended a seminar held by the Kocaeli branch of the Junior Chamber International at Sabancı Culture Center, Kocaeli, from December 12 to December 13 in 2009. The study found that the top three pull factors in students' intention to migrate were 'great number of professional advancement opportunities', 'high living standard', and 'proximity to great science and innovation centers'. The top three push factors were 'inability to develop oneself at a better level in his/her field', 'inadequate opportunities in professional advancement', and 'economic instability'. 79.5 percent of the participants reported that they have intention to continue their education abroad, 73.8 percent intended to work abroad, and 53.3 percent would like to live abroad.

Moreover, Akman (2014) found that the pull factors (i.e., high living standard, social life, culture, health-care services, great number of professional advancement opportunities, deferment of military service, economic stability, systematic well-ordered environment, opportunity for children to get better education abroad, proximity to great science and innovation centers) in male students' intention to migrate were statistically more important than in female students' intention. The male students were found more likely to intend to work and settle abroad. Younger students were found to be more likely to prefer to continue their education abroad. His analysis suggests that pull factors are more influential in staying abroad for longer periods. Also, according to the findings of the study it is safe to conclude that the perception that there are more job opportunities abroad than in Turkey is a key motivating factor in emigration decisions. Push factors (i.e., low salaries, inadequate opportunities in professional advancement, economic instability, political instability, healthcare services, corruption, bureaucratic barriers, ill-functioning government agencies, inability to develop oneself at a better level in his/her field, distance from great science centers in the field, lack of material support and financing required for setting up a business, unsatisfactory social and cultural life, and inadequate social security) are influential in the decision to live abroad.

Mollahaliloğlu et al.'s (2014) study is conducted in 2009, based on 3690 out of total 4753 physicians who recently graduated in Turkey. The participants, on a voluntary base, completed the questionnaire during their job application to the Ministry of Health. The authors employ a logistic regression method. Although 54.6 percent of the participants indicated that they thought of pursuing graduate education (specialty/PhD) abroad, about 80 percent of them did not make any attempt on this regard. For about 70 percent of the participants, the most important factor in considering living (work or study) abroad is 'working conditions' (p. 71). The study found that the desire of male physicians is 1.5 more than that of female counterparts, and it is statistically significant. The study also revealed a highly strong correlation between foreign language skill level and willingness to reside abroad. The study, however, found no significant linkage between the development level of cities the participants were born, the income group of their parents, and their willingness to reside abroad (p. 72).

- *The second group: targets returnees*

Altaş et al. (2006) focus on currently employed 200 academicians from 27 universities in Turkey who returned from abroad. The study aims at investigating the possible relationship between their migration and return decisions by utilizing a loglinear model. The study does not reveal a statistically significant relationship between primary factors of migration and return decisions. While 'missing family/home country' was the main reason in the return decision for those who were funded by their university (i.e., teaching assistants or research assistants), it was 'compulsory services' for those who were not funded by their universities (i.e., those with MoE scholarships). For both groups, however, the second most important factor was the desire to apply/carry on attained knowledge/training in home country.

Vatansever Deviren and Daşkıran (2014) aim to examine the decisions of faculty members of Muğla Sıtkı Koçman University in terms of their choice of university, return decisions, satisfaction with their decision, and their opinion on the brain drain focusing on those who had earned a degree from a foreign university or spend some time for academic purposes in a foreign university. To this end, the authors reached out to 106 academicians with a survey of 27 questions. 43.4 percent of the participants had been in the US, 18.9 percent in the UK, and 13.2 percent in Germany, followed by other countries with very small proportions. The survey revealed that while 47.1 percent of participants are happy with their return decision, the rest is either not happy (18.9 percent) or undecided (34 percent). Although the study only covers a single university, this is an important input for policymakers. According to the survey, the main reason for the return decision is the compulsory services (67.9 percent). Desire to utilize their attained knowledge and training in home country was another highly ranked response with 65 percent. Better social and cultural environment in Turkey (43.4 percent) and relatively easy professional promotion in Turkey (40.6 percent) were two other popular responses. About 69 percent of participants who expressed their desire to migrate indicated that they would like to return abroad to involve in research projects or other research and development activities. In the survey those who expressed their desire to migrate were also asked to make some suggestions to prevent brain drain. Accordingly, harmonization of working conditions with abroad, improving the library facilities, devoting more funds for scientific activities, improving the conditions of lab in natural sciences, increasing pay, improving technological facilities, reducing teaching load, having long-term plan for science and technology, encouraging interdisciplinary studies, increasing the number of independent research institutes, making a clear separation of research and teaching universities as has been implemented abroad, and extending academic freedom were stated by the participant faculty members (Vatansever Deviren and Daşkıran 2014).

- *The third group: targets stayers and returnees*[6]

Uysal (1972) is reviewed in Başaran (1972a, b). The study was funded by the Turkish Scientific and Technical Research Institute (TÜBİTAK) and conducted in 1972 by associate professor Şefik Uysal. Dr. Uysal reached out to scholars who were sent abroad or were currently in abroad in accordance with Law No. 1416 between 1929 and 1972[7] (Uysal 1972). Among those who returned to the country, the reasons for leaving the institutions were the lack of value given to specialization (47 percent), working on a project not in the related field (30 percent), and insufficient salary (23 percent).

Köser-Akçapar (2006, 2009) aims at discussing the 'brain drain/gain' issue for Turkey based on mixed qualitative-quantitative data to answer whether 'brain drain' is detrimental for Turkey and to examine the costs and benefits of the phenomenon. The study is based on in-depth interviews conducted with 140 people[8] selected according to their departments and working sectors (p. 6).

Köser-Akçapar (2006) shows that the Turkish graduate students mainly reported that 'adaptation problems due to cultural differences, missing Turkish food, loneliness, homesickness and being away from family and loved ones, F1 visa problems, racism and discrimination in some cities especially after 9/11, and some financial problems' were major problems they faced (p. 10). However, the key point is that 46 percent of the respondents had not encountered any problems in the US at all. Although only 16 percent of the respondents expressed that they would like to reside in the US, 42 percent noted that they were undecided.

A unique question that was asked to the participants in Köser-Akçapar (2006) is whether the harmonization of policies with the EU would affect their decisions to return home after completing studies in the US. Accordingly, while 62 percent expressed that it would, 26 percent said maybe, and 12 percent noted that it would have no effect because either they 'were planning to return anyway, or they made up their minds not to return' (p. 11). Among young professionals, there were no complaints about wage discrimination but the respondents noted the existence of a glass ceiling and the difficulty of finding jobs as easily as it used to be before 9/11 (p. 11). The study found that many Turkish students do not return right after finishing their studies, but instead they prefer to stay and work. Forty-seven percent of the respondents reported that they are permanent in the US, whereas 35 percent of them were not sure, and 18 percent of them noted that they would definitely return within a couple of years. Although the study does not indicate the rankings clearly, it reports that economic reasons (i.e., wage differentials, higher living standards, unemployment, underemployment), personal reasons (anxiety about the future, children's education), political reasons (political instability, bureaucratic obstacles, corruption), and professional reasons (not enough R&D, lack of scientific research at universities, lack of opportunities for highly skilled studies in Turkey) are reasons for not to return cited by the respondents. The study reports that among the respondents whose age ranged from 47 to 77, 92 percent of them said that they consider themselves staying permanently in the US.

Another crucial finding of Köser-Akçapar (2006) is that although 10 (out of 25) respondents returned to Turkey for some time ago, they remigrated to the US because of a set of problems they encountered when they were in Turkey. The study also reveals some findings based on ten respondents who returned to Turkey after finishing their study. In addition to a combination of 'personal and professional reasons', 'the developments and economic improvement in Turkey' and 'the difficult environment after 9/11' were also cited by respondents as the main factors for returning decision. Finally, the study reports that all of the ten undergraduate students from different fields reported that they would like to 'return to Turkey after completing their graduate studies and pursue careers in Turkey' (Köser-Akçapar 2006, p. 14).

In a recent doctoral dissertation, Esen (2014) investigates the determinants of the decision of Turkish PhD holders educated in the US to return to Turkey or to stay in the US. Esen (2014) refers to a recent study that showed that there are mainly two reasons why Turkish government-sponsored research assistants chose to stay in the US: first, their view that less value was given to academia in general and to science in particular in Turkey and second, their concern about the lack of productivity in the academic environment in Turkey. Some of them went as far as to note that 'if science and academia were respected and valued, they would not care about the fact that wages are lower in Turkey' (Çelik 2012, cited in Esen 2014, p. 2).

Arguing that there has been improvement in social and economic conditions in Turkey since 2003, Esen (2014) aims at investigating how these 'better' conditions affect educational mobility toward an economically rapidly developing home country like Turkey. Esen also attempts to investigate two crucial questions. First is whether the US-educated Turkish scholars actually have the option to stay in the US or whether they return because they are not hired. The second question is what are the perceptions of these scholars with respect to the initiatives/attempts by the Turkish private and public institutions and availability of the European Union funding entities.

Esen (2014) prefers a qualitative approach to investigate the brain drain issue. To this end, he reaches out to 20 US-educated scholars teaching in universities in the US or in Turkey. Esen notes that virtually for all returnees the role of the economic crisis in the US was not a major factor in their return decision. Esen summarizes the perception of these scholars toward the initiatives by TÜBİTAK: 'good idea but problematic incentive' (p. 54). The primary reason for return was 'family ties, cultural values, and social difficulty', while only one participant mentioned the positive role of the economic growth in Turkey (p. 57). Except for one participant, the stayers felt uncomfortable with the current insufficiency of the academic situation in Turkey. The majority of the stayers valued research opportunities more than wage opportunities. In fact, a participant noted that 'when salary and fringe benefits are considered, the Turkish private universities are a better deal' (p. 68). They noted the lack of good academic environment, the presence of bureaucracy, nepotism, and interference of politics in academia; 'politics dominates all aspects of life' (p. 71). Participants noted that 'academic freedom, political stability, and academic opportunities would be more important reasons than economic growth for them to ever decide to go back to Turkey' (p. 72).

Scholars interviewed in Esen (2014) also highlighted the poor economic performance in the US. However, when it came to the decision to return to Turkey, they revealed the same concerns about the higher education system and academia in Turkey. Esen (2014) concludes that '[a]lthough Turkey is experiencing some positive changes, such as economic growth and political stability, which could enable the government to offer more financial incentives to U.S. educated Turkish scholars, some participants did not seem to be attracted by those opportunities. This study found that the majority of the participants preferred to stay in the U.S. as a result of the greater availability of academic opportunities at universities there' (p. 76). Esen (2014), in line with Çelik (2012), emphasizes the returnees' dissatisfaction with work conditions and environment in Turkey. Those scholars 'complain about not receiving enough support from their employers, which limits their ability to apply their knowledge and experience', and Esen states what his study confirms and emphasizes this as a discouraging factor in returning (p. 87).

- *The fourth group: focuses on students and/or professionals abroad/stayers*

To the best of our knowledge the first empirical examination of the brain drain in Turkey is a joint study on medical doctors conducted by Ankara Hıfzısıhha School and John Hopkins School of Public Health dated 1968 (Dirican et al. 1968, cited in Başaran 1972a, b). The study is based on 230 doctors living abroad, consisting of 18.3 percent of Turkish doctors in the sample. According to this study, by far the first reason for immigration (68 percent) was insufficient income, followed by the difficulty of professional promotion (12 percent), the difficulties faced in fulfilling the role of being a doctor (6 percent), the desire to live comfortably (6 percent), and inadequacies in the organization of health services (6 percent). The study states that one-fifth of these doctors had already intention to go abroad when they were in Turkey (Başaran 1972a, b).

Kösemen (1968) is reviewed by Başaran (1972a, b). It is a study on engineers prepared by Cevdet Kösemen in 1968. According to Kösemen (1968), in 1968 there were 975 Turkish engineers and architects working abroad, which compromised the 5.6 percent of all engineers and architects in Turkey (Başaran 1972a, p. 66). The study suggests two clear, perhaps not unexpected, results in that (i) those who had their degrees from a university with medium of English language in Turkey were more likely to migrate than other university graduates and (ii) those who earned their degrees abroad also were much more likely to end up with abroad. Lower wages in Turkey, more possibilities to learn foreign languages, desire to see different places, hierarchical authority and political pressure, and various frustrations in working places were main reasons for brain drain among engineers.

The research conducted by Oğuzkan (1971), which is funded by the Turkish Scientific and Technical Research Institute (TÜBİTAK), is another early study to understand the causes of brain drain (Oğuzkan 1971, 1975, 1976). Between 1956 and 1970, 907 Turkish engineers and 594 Turkish medical doctors came to the US (Oğuzkan 1976). According to UN, between 1962 and 1967, 375 Turkish scientists migrated abroad, 51.1 percent were medical doctors, 40 percent were engineers, 5.5 percent were scientists, and 3 percent were social scientists (Erkal 1980, p. 80).

Oğuzkan (1971) surveyed 150 scholars holding PhD degrees. The findings of the study showed that the most important attracting factor was 'professional' issues, namely, the possibility of finding a job in the field of specialization, availability of financial and other means for work, the opportunity for professional advancement, the closer contact with important scientific centers, and so on. In addition, the study showed that economic and sociocultural reasons were important reasons in decision-making process as well. The study provided very similar results in terms of push factors. Oğuzkan's study also revealed two crucial facts. First, regardless of their intention to return to Turkey, scholars indicated that they need to follow the recent developments in Turkey, which show the possibility of important linkage between home and abroad. Second, the findings very clearly showed that those who earned their degrees from abroad were much more likely to reside abroad. Some discussion on the findings of this study can be found in Atabek (1971) and Tezcan (1971), where the authors conclude that the physicians have complaints about the general instability (insecurity about future) and working conditions. The physicians emphasized the overall increase in their welfare level in the US, the availability of the latest medical technology, and satisfaction with professional development (Tezcan 1971, p. 57).

Yavuzer (2000)'s study is based on a survey of 363 individuals as well as interviews and personal observations conducted in Washington D.C., Maryland, and Virginia. The study confirms some expected results such as the finding of a higher education level of the Turkish nationals residing in the US compared to those living in Turkey or Europe. Also, it confirms the quick adaptation process for individuals who have graduated from universities like Middle East Technical University, Bilkent University, and Bosphorus University. Regarding the return intentions, while 59 percent of participants intended to return soon, after graduation, or after 5–10 years, 41 percent intended either to return in retirement or not to return at all. However, the author emphasizes that it is more likely that the reported return intentions are more wishful thinking than realistic statements (p. 101). He notes that based on personal interviews many participants noted some various difficulties in returning (i.e., business/personal ties in the US). Finally, the author notes some general information taken from the Turkish Embassy in that graduates of the aforementioned top universities in Turkey compared to other university graduates and students of natural science compared to other fields are more likely not to return.

Öztürk's (2001) study is one of the most comprehensive studies that attempt to understand the motivations behind going abroad and factors that affect the return decision of the Turkish students residing in the US. The study is based on the data collected via questionnaire and in-depth interviews with 330 students between 1997 and 1998. The authors utilized one-tailed t-tests and a regression analysis to examine the data.

Öztürk (2001), in terms of students' motivation to study abroad, found most of the external motivational factors important. While the majority of students reported that 'better work opportunities abroad' and 'immigration to the host country' were not important, 'availability of better educational facilities outside Turkey', 'the quest for freedom', 'availability of scholarship to study abroad', 'educational diversity of the programs abroad', and 'a broader understanding of the world' were important factors. In terms of internal motivational factors, the majority of students reported that 'avoiding mandatory military service in Turkey', 'suitable marriage', 'political uncertainty in Turkey', 'non-acceptance at a university or college in Turkey', 'the absence of their field of study in Turkey', and 'the satisfaction of parents' were not important. On the other hand, 'the prestige and better opportunities at home with a foreign degree', 'the availability of scholarship to study abroad', and 'being able to serve to Turkey in the future' were important.

Öztürk's (2001) study also provides the US-specific motivational factors. Accordingly, while 'influence by American movies', 'immigration to the host country', and 'influential American missionary schools in Turkey' were found to be not important, 'the prestige and better opportunities at home with a US degree', 'availability of scholarship to study in the US', 'educational diversity of the programs in the U.S. higher education institutions', 'availability of superior training facilities in the US', 'greater choice of American higher education institutions', 'the availability of advanced research facilities in the US', and 'the possibility of academic research and job opportunity' were important factors that encourage students to choose the US over another country. Moreover, 'financial assistance', 'financial aid information', 'cost of higher education', 'racism in Europe', 'reputation and image of the US', 'business/job opportunities', 'sponsor's recommendation/requirements', 'educational facilities', 'language', 'networking', 'American domination in publications', 'advanced science/technology', 'dynamic education system and educational diversity', 'admiration and closeness to the culture', and 'cultural diversity' were effective in students' motivation to choose and study at an American university over a European university.

Öztürk's (2001) study shows that while 44 percent of the students have the intention to return to Turkey immediately after completing their studies in the US, about 30 percent of them prefer to stay in the US, and 26 percent were undecided.

Öztürk (2001) found that the following hypotheses were statistically significant (pp. 214–220): 'Privately-subsidized students are more likely to stay in the US after graduating than the Turkish government sponsored students'; 'The higher level of funding Turkish students receive from the Turkish government, the lower the intent to remain in the US'; and 'Turkish students who have higher socio-economic status background are more likely to intent to remain in the US'.

The following hypotheses, however, were not statistically confirmed to be true (Öztürk 2001, pp. 214–220): 'Female students are more likely to return to Turkey after graduating than male students do'; 'Turkish students who are married to Turkish citizens are more likely to return to Turkey than single students are'; 'Turkish students with higher level of private schooling (prior to their college education) are less likely to intend to return to Turkey after graduation'; and 'Turkish students for whom the prestige and better opportunities at home with a US degree as motivation for overseas education has a higher level of importance are more likely to return to Turkey after graduation'.

Öztürk (2001, pp. 270–272) revealed that 'being among friends, relatives and family' is the most important personal/cultural advantage for students to return to Turkey, followed by 'the possession of the knowledge of English language' and 'a broad understanding of the world'. 'Bringing knowledge and advanced technology to Turkey' is the most important national advantage factor for students to return to Turkey, followed by 'serving Turkey' and 'the growth and competitiveness in Turkey'. In terms of professional advantage factors of returning to Turkey, 'availability of better job opportunities', 'to be better educated than other Turkish citizens', and 'a faster job development and promotion opportunity' were the most factors.

On the other hand, Öztürk (2001, pp. 270–272) reveals that 'life standards and conditions of Turkey' and 'freedom of expression and the political instability in Turkey' were two most important personal/cultural disadvantages of returning to Turkey. Finally, 'unsuitable environment for professional growth', 'potential low-income opportunities and economic difficulties due to unsatisfactory salaries', and 'lack of opportunities to fully utilize experience gained abroad' were found to be the most important professional disadvantage factors for students to return to Turkey.

Gökbayrak's (2009) work is based on a questionnaire that was distributed between the end of 2004 and early in 2005. The author preferred open-ended questions to enable respondents to explain their opinions in detail on their connections with Turkey and colleagues in Turkey, participation in joint projects and their evaluation, and their intentions to support Turkey. The study also includes some closed-ended questions relating to push and pull factors related to migration and characteristics of living and working abroad. The author notes that a key difference between her work and the previous works in the literature is that the respondents in her work are only those who had become permanent residents in foreign countries (p. 138).

Gökbayrak's (2009) study is based on over 130 responses returned by engineers, which are analyzed by Chi-square tests and correlation analysis. The findings are discussed in light of additional information based on other data collected and in-depth interviews. The study also points to a correlation between being from universities with medium of education in English and being in abroad. The study finds that the most important factors in migration decision were dissatisfaction with professional environments regarding further specialization in the field and dissatisfaction with pay and human relations. The study reveals similar results in the case of pull factors.

In line with other studies, Gökbayrak (2009) also finds that there is a very strong tie between scholars at home and abroad. 83.5 percent of participants reported that they kept in touch with their colleagues in Turkey. However, in case of the collaboration in professional works the ratio is as low as 23.8 percent. The author also notes that the respondents expressed their dissatisfaction with these ongoing joint works as a result of tensions in human relations and bureaucratic obstacles (p. 143). As an important indicator of financial personal ties with Turkey, the study shows that less than 5 percent of the respondents had investment in Turkey.

Gökbayrak (2009), along with the previous studies, showed the existence of the tendency not to return over time. While 77.8 percent of the respondents considered returning to Turkey when they had first settled, at the present time of the survey 52 percent of them had tendency to return. The findings of Gökbayrak (2009) in terms of present intentions to return of the respondents (as high as 52 percent) are contrary to the findings of previous studies of Kurtuluş (1999), Öztürk (2001), and Güngör and Tansel (2008a). Following up the statement that females generally have a weaker tendency to return compared to males (Tansel and Güngör 2004), the study also pays attention to the possible differentiation in terms of gender. However, the findings do not provide statistically significant results in terms of the impact of gender in return decision due to the limited number of female respondents. While it is economic and political instability that was the leading factor in (not) return decision in Tansel and Güngör (2004), it was second most important reason (after professional concern about the further specialization) in Gökbayrak's study. This is, as the author notes as well, because the former study was conducted right after the 2001 crisis in Turkey. Gökbayrak (2009) notes that the impact of age and original decision on actual return is similar to the findings of Tansel and Güngör (2004). Besides improved professional environments, participants working abroad expected political and economic stability in Turkey. The third expectation was a higher-quality socio-economic life. This broad category refers to improvements in many fields including social security, health, education, transportation, and public security. Gökbayrak (2009) states that in order to return the respondents seek for improved professional environments, political and economic stability in Turkey, and a higher-quality socio-economic life (i.e., extended welfare state). Although the present ties are not strong with colleagues in Turkey, the respondents expressed their willingness to expand the ties.

Pazarcık (2010), in her study, chose top 54 American universities from the *Times Higher Education's* 'The World's Best Universities' ranking (www.timeshighereducation.co.uk). The study is based on surveys completed by 50 Turkish academicians from those 54 universities. More than half of the participants were born in İstanbul, Ankara, or İzmir. The study reveals some descriptive statistics that provides some hints about the potential of return intentions of the participants. For example, 40 percent of spouses are the citizens of foreign countries, of which more than 80 percent are US citizens (p. 62). Thirty-six percent of them have some sort of assets or real estates in Turkey (pp. 65–66). Fifty-four percent of the participants used to work in Turkey before moving to the US, of which 63 percent used to be academicians (p. 69).

Pazarcık (2010) reveals that the most important push factors for the brain drain were the lack of academic merit, dissatisfaction in work place/not suitable environment for specialization, the lack of academic/scientific infrastructure (i.e., labs, etc.), and lower wages in order of importance. The least important factors in order were stated as dissatisfaction with the social life, unemployment, and political reasons. Pazarcık (2010) shows that the most important pull factors were availability of the best program in one's specialization, higher pay, applicability of academic training, and peaceful working environment in order of importance. Eighty percent of the participants think that the policies in Turkey lead to the brain drain

(p. 79). In the same manner 76 percent of participants think that there exist encouraging policies to pull brain drain toward the host country they reside. Eighty-two percent of the participants are satisfied with the pay level, and over 90 percent thinks there is no discrimination in the work place or in social life including wage discrimination. Sixty-two percent of the participants noted that they pursue an academic relationship/work with their colleagues in Turkey. However, only 44 percent of participants have been involved in joint work with their colleagues in Turkey. The three most important factors for return decision were improving salary, improving availability of funds for research activities, and improving academic autonomy. In a similar manner, the participants noted the advantages of the academic environment compared to Turkey. Accordingly, availability of more funds for research, higher salaries, more intellectual and skilled colleagues, academic freedom, and appreciation of academic success are most referred facts. Pazarcık's (2010) findings show that while 86 percent of participants used to think that they would return to Turkey when they first arrive, now 64 percent of the participants think that they do not consider returning to Turkey, which is a very clear finding.

3.2.1 Works of A. Tansel and N. Güngör

It should be noted that there are several versions of works by Aysıt Tansel and Nil Demet Güngör, based on N. Güngör's doctoral dissertation (Güngör 2003), such as Tansel and Güngör (2002, 2003, 2004, 2009). However, here we would like to cover the three most recent main publications (Güngör and Tansel 2008a, b, 2014). To the best of our knowledge Güngör (2003) is the most comprehensive empirical work on the brain drain in Turkey. Reaching out to more than 2000 people and presenting excellent analyses, the study provided inspiration for this project and became the departure point for it. That is, the main goal of this project was to provide some recent findings to compare and contrast with those of Güngör (2003). That is why the questionnaire used in this study is adopted from Güngör (2003).

Güngör and Tansel (2008a) note that according to informal evidence compared three decades ago, there are more individuals with higher education emigrating from Turkey today (p. 3070). The study is based on data obtained through an Internet survey conducted by the authors during the first half of 2002. The authors utilized an ordered probit model of return intentions of Turkish students. The dependent variable is the return intentions of the students, regressed on several independent variables.

Previous empirical works showed that women have higher tendency to stay in the resident country. The study finds that female students have less tendency to return to Turkey, perhaps due to the gender gap in the labor market and less freedom in social life (ibid., p. 3073). While this finding is statistically significant in simple specifications with few variables, the final preferred model does not present significant results.

Güngör and Tansel (2008a) found that 'the probability of having weak return intentions increases with the age of the participant at a decreasing rate' (p. 3073). The authors argue that this might be the case as it might be more difficulty for older students to return because of its greater psychic costs for them. Once again age variables were only significant in the simple models with few variables.

The findings reveal a statistically significant relationship between the length of stay in host country and return intentions. Accordingly, first, strong return intentions decline as the time spent in the host country increases; second, those who report strong non-return intentions become much more persistent in their decision as the time passes. The study also provides evidence that initial intentions to return are a key factor of return intentions. It is so, because for those who compare the less favorable conditions in home country and greater opportunities in abroad consider study/work abroad as an opportunity to escape from the home country. Also, as expected, the study reveals that greater family support and being married to a foreign spouse increase the probability of not returning.

The authors argue that the role of socio-economic background is more important 'in determining who can take advantage of study opportunities abroad than in determining students' return intentions once abroad' (p. 3078). This is also true in the case of the prior knowledge of foreign language. Social life and standard of living are also two significant factors in return decisions. The study does not reveal that the return decisions vary with respect to the field of study significantly.

The study presents some more results, which are also not unexpected. Accordingly, while, the probability of return decision is higher for those who are abroad as a job requirement in Turkey or have compulsory academic service, it is smaller for those who went abroad because of the political environment in Turkey, lifestyle preferences, or lack of sufficient equipment for research in Turkey and for those who had previous experience in abroad.

Güngör and Tansel (2008b) note the very high education level of Turkish emigrants in the US. The authors refer to the 2000 US population census stating that while the proportion of those holding an MA degree or above is 12 percent among the Armenians, and 10 percent among the Greeks, it is as high as 23 percent among the Turks. The study is based on 1224 responses obtained through an Internet survey conducted by the authors during the first half of 2002. The survey universe is the Turkish scholars and professionals working at a full-time job abroad and holding a tertiary-level degree.

The authors state that since higher education system in Turkey cannot respond to the demand for higher education, some students prefer to pursue their education abroad. A large part of this group is private students who are financed by their families. Another important section is those who receive scholarships from different government bodies (p. 325). As an expected result, 'the respondents come from relatively well-educated and well-to-do families who were able to invest in the higher levels of education in Turkey' (p. 328).

Güngör and Tansel (2008b) show that more than half of the survey participants (55.4 percent) have graduated from high schools whose medium of language is not Turkish. Also, in line with other studies, a majority of respondents have BA/BS degree from universities whose medium of language is English, such as Middle East Technical University, Bosphorus University, and Bilkent University. Also, a nonnegligible part (11.5 percent) of respondents holds an undergraduate degree from foreign universities. The study reveals a significant positive relationship between initial and current return intentions. Eighteen percent of respondents reported that taking advantage of educational opportunities is the most important reason for going abroad. The need for change, lifestyle preference, and the lack of facilities and necessary equipment for carrying out research in Turkey were other important factors in order of importance. However, the top reasons for going abroad differ with respect to the highest degree completed by respondents. Accordingly, while for bachelor's and master's degree holders it is the need for change and lifestyle factors, it is research-related factors for PhDs.

Moreover, Güngör and Tansel (2008b) found that gender also matters. 'Female respondents are more constrained by family rather than individual considerations' (p. 333). Respondents report that economic instability is the top reason for not returning. As the authors noted as well, this is not unexpected as the survey took place right after the 2001 crisis in Turkey when the rate of unemployment among the youth reached peak levels. The study shows that other main reasons are bureaucracy, unsatisfactory income levels, political instability, and lack of opportunities for advancing in occupation in order of importance. On the other hand, high salaries and more organized/ordered environment were two top reasons as pull factors.

Güngör and Tansel (2008b) reveal that a common concern about academic career is the lack of value given to science and to academics in Turkey. Through a correspondence analysis it is shown clearly that initial intentions have a positive relationship with current return intentions, and this positive linkage weakens with the length of stay. Also, it is shown that previous work experience is an important determinant of return intentions. The study did not find significant association between the R&D intensity of job activities and return intentions.

Güngör and Tansel (2008b) show that those who are most likely to report not wanting to return to Turkey are those who 'returned to Turkey to work at a full-time job immediately after completing their studies at a foreign university and who then decided to go abroad again to work'. On the other hand, those who have obtained their highest degree from a Turkish university are the group that is more likely to return to Turkey. On the other hand, respondents with a foreign highest degree, regardless of level, are least likely to return, and those who earned their highest degrees from the Turkish universities have the definite return plans.

Güngör and Tansel (2014), as a complement study to Güngör and Tansel (2008b), found that the income difference between home and host country is an important consideration for the majority of emigrants. It is the initial intention not to return that has the greatest positive impact on the probability of not returning. The study also shows significant impact of family support to settle abroad and marriage to a foreign spouse, longer stay duration, work experience, and specialized training abroad. As the study based on the survey conducted right after the 2001 economic crisis in Turkey, it found that economic instability in Turkey has the greatest deterrent effect on return intentions. Also, compared to other professionals, academicians are more likely to stay abroad. The study found that there is a clear difference in tendency for females to report their intention to stay abroad compared to males. The authors argue that this is because they are more inclined to enjoy the greater freedom of lifestyle abroad and in line with previous findings in the literature for China (Zweig and Changgui 1995).

3.3 Current Debates and Reverse Brain Drain

In the recent years there has been a significant debate about increasing intentions of highly educated individuals to leave the country due to political instability accompanied by major initiatives, such as handsome scholarships, to reverse the brain drain.

The phenomenon, so-called reverse brain drain, has taken a place in the agenda of the policymakers in some developing countries including Turkey. This is a result of economic crises in the advanced countries on the one hand, and persistent economic growth in home countries on the other hand. In this context, increasing investment for research and development activities and some financial incentives in Turkey have made 'home country' attractive.

There exist three policies of implementing brain gain: (1) governments may try to retain their students; (2) encourage students to pursue postsecondary education abroad and return, and (3) engage with the diasporas (Papila 2015). China, South Korea, Taiwan, India, Pakistan, Mexico, and South Africa are major examples that pursue these policies. In this context, although there is a strong intention to reduce the brain drain and create a reverse brain drain, there is no solid policy on these issues in Turkey. However, there are still important implementations worth mentioning. For instance, İçduygu (2009) argues that the establishment of private universities and competitive facilities attracts Turkish scholars, scientists, and university graduates living abroad back to the home. More importantly, the government bodies put emphasis on the research and development projects in general and reverse brain drain in particular. In this context, TÜBİTAK, as the main science and technology institute, prepared 'the Science and Technology Human Resources Strategy and Action Plan 2011–2016' (TÜBİTAK 2010). With this, the share of funds allocated from Turkey's GDP to R&D investments has doubled from 2002 to 2012 (from 0.5 percent to 1 percent). However, it is worth noting that this ratio is still lower than the OECD average.

In the case of Turkey, there are several channels to promote the return of students, namely, TÜBİTAK, Turkish Aerospace Industries, and the European Union. EU-funded projects, for example, the Marie S. Curie Reintegration Grants, are one of them. During the 2007–2014 period, Turkey received 204 Marie Curie Career Reintegration Grants. In this regard, thanks to TÜBİTAK's 'reverse brain drain' implementations, 595 researchers returned to Turkey in the last eight years (235 through Marie

Curie scholars and other supports by European Research Council Program and 360 through national funds provided by BİDEB).

TÜBİTAK has also initiated a reverse brain drain campaign to invite Turkish researchers who are in the US to return to Turkey permanently. That campaign was supported by the European Council. Since 2010 TÜBİTAK has had 12 'Target Turkey Workshops' in different places in the US and Europe, which reached out to about 1500 Turkish scholars (TÜBİTAK 2015). Esen (2014) provides a valuable summary of the different reactions that emerged in these meetings. While positive views emphasized the increasing opportunities made possible by increasing the amount of funds provided by different institutions, others emphasized the lack of freedom of expression, the nonautonomous structure of higher education in Turkey (e.g., Yüksek Öğretim Kurumu), and the lack of a solid science policy.

Since its establishment in 2010, BİDEB supported 360 researches out of 597 applications, and by April 2015, there were 67 applications that were being reviewed. Moreover, TÜBİTAK provided 1323 grants out of 5491 applications in 2012, and the number of applicants jumped to a historic high level of 15,030 in 2013. It offered scholarships of TL 3250 ($1475) per month for a duration of two years. TÜBİTAK also offered grants of up to TL 25,000 ($11,351) for research to be conducted in Turkey, which also included incentives such as health insurance and a plane ticket to Turkey (TÜBİTAK 2015). To promote brain gain more than 83 fellowships were provided by TÜBİTAK-BİDEB between 2012 and 2014 (Papila 2015). Moreover, the Turkish government started appointing 'scientific ambassadors' in the US, EU, and Japan in 2014 to provide assistance to the needs of the Turkish scientific diaspora, which is another important development in the context of brain gain (Papila 2015).

Akyol (2014) notes that although those who intend to return to Turkey cite 'the perception of a better work environment, the increasing provision of resources and investments, and TÜBİTAK's contributions and EU funding programs', there are still serious concerns including but not limited to 'too much bureaucracy and nepotism'. The author also states that the teaching load is another concern of potential students who planned to return. Akyol (2014, p. 1) argues that '[i]t is the political crisis and the political culture that remain justifiably worrisome for many, both inside and outside Turkey'.

Güler (2009) and Ünver (2010) note an increase in reverse brain drain in Turkey. However, Ünver (2010) emphasizes that it is not the convenient conditions in Turkey but economic and social problems such as xenophobia and discrimination abroad that are causing the reverse brain drain (Ünver 2010, cited in Bilgili and Siegel 2011, p. 24). There have been implementations toward facilitating the permissions for the employment of foreigners in Turkey and cutting down the red tape process to reduce the length of the process (Kurtsal 2011). Also, there are some initiatives to encourage successful foreign students to keep residing in Turkey after graduating. Moreover, the law of 4691 of the Technology Development Zones and the law of 5746 of Supporting Research and Development Activities are two of the legislative implementations that serve to prevent the brain drain (Kurtsal 2011). In this context, according to Article 31 of the Law on Foreigners and International Protection, adopted in 2013, a short-term residence permit can be issued to a foreign student with a maximum validity of one year.

Papila (2015, p. 1) argues that 'while constructing a functional bridge between the scientists/researchers in the US and Turkey, we should consider a two-way flow of skill, capital, and technology and focus on brain circulation rather than reverse brain drain as it is a more preferable option among the Turkish Science Diaspora in the US'.

Papila (2015) suggests that, first, EU funds can be used to facilitate brain circulation. The author suggests that a scholar in the US can apply for this grant to set up its team and lab in Turkey for half of his/her time while still keeping her/his position in the US. Second, entrepreneurship activities can be extended to Turkey. Technology Transfer Accelerator Turkey (TTA Turkey), designed by the European Investment Fund (EIF), aims at achieving two goals: '(i) setting-up a financially sustainable fund by facilitating the commercialization of scientific research and development (R&D) confined in universities and research centers; and (ii) catalyzing development of the technology transfer market in Turkey, with a particular emphasis on spill-overs to the less developed/developing regions of Turkey'.

Tanyıldız et al. (2011) provide some policy recommendations. Accordingly, (i) there should be reforms/initiatives in regulations and patent and intellectual property rights in order to encourage returns from abroad; (ii) both scholars and staff in innovative sector should be provided special incentives; (iii) local administrations should shoulder the responsibility of encouraging skilled workers and provide incentives; (iv) government should encourage junior scholars to study in appointed fields so that on the one hand companies may reach the skilled workers they demand and scholars may have job guarantees when they return after completing their studies; (v) there should be linkage between students' study fields abroad and demand by industry; (vi) the government should encourage skilled labor not just at the stage of innovation but also at marketing stage as well; (vii) creative finance methods should be developed; as in the case of Israel, the innovative sectors should be supported by tax incentives or with similar mechanisms in economic recessions as well; and finally, (viii) foreign scholars should be incentivized to migrate to Turkey.

The recent social/political environment has had a substantial negative impact on the aforementioned initiatives toward reversing the brain drain. Although there are no academic works dealing with this effect yet, it has been addressed by the press. There has been an ongoing debate over an increasing brain drain due to political instability. Kınıklıoğlu (2014), former Justice and Development Party (AKP) deputy, reports his growing concern about brain drain as educated people are frustrated with the growing authoritarianism and chaos in the country. He reports, based on his personal observations, a growing number of people asking for emigrating to foreign countries. He notes that

> [w]hite-collar Turks employed in international companies are seeking out positions in the overseas offices of their companies. Those who have the financial means are looking for opportunities to buy a home or start a business abroad as a way to legally settle down there. Some are going to relatives based abroad, trying to join them there. Others are going abroad on temporary job contracts and then scrambling to get permanent positions. In the United States, the number of green card applications [by Turks] has significantly increased. Put simply, white-collar Turks are desperate and leaving. They are looking for a better future. They don't want to raise their children in such a country.

He also refers to relevant research that indicates that 'political instability and authoritarian rule lead people to lose trust in their government and state'.

Kınıklıoğlu (2014) notes that a reverse brain drain in the 2007–2011 is no more the case, and Turkey once again—just like it happened in the 1970s, 1980s, and 1990s—started to lose human capital. He goes on further to note that '[i]n previous years, the flights were driven mostly by economic and opportunity-related reasons, while today the motive is mostly ideological. The human capital flight is being further accelerated by the increased fanning of social polarization and alienation in recent years.'

Kınıklıoğlu's (2014) observations are supported by a research conducted by Globalcv.com, an online job portal which receives more than 100,000 resumes a month. The research states that 'many white-collar workers are discriminated against because of their lifestyle, worldview and political ideology' (FactsonTurkey 2014). The company received numerous applications from high-level professional Turks living in the US who wanted to return to Turkey before 2012. However, they note that the trend has reversed with the increasing polarization in the society.

Moreover, a news in *Amerika'nın Sesi* [Voice of America] based on interviews with the emigration lawyers located in the US, and the president of the Turkish-American Chamber of Commerce and Industry, reports the boom in demand for migrating to the US in 2016 (Kamiloğlu 2016). An emigration lawyer said that there has been a boom in demand for settling in the US in the last seven to eight months. The news reports that there are even demands from some celebrities in Turkey. Most of these demands are coming from higher-income groups, who are able to start a business in the US. Another group who has increasingly requested information to migrate to the US are those who used to live in the US but for some reason decided to permanently return to Turkey. Those who would like to reside in the US express their concern about the increasing terror and their concern about the education of their children. There has been a boom in demand for investing in the US by Turkish business people as they are extensively frustrated by the increasing instability (Kamiloğlu 2016).

In addition, on the same concern, Aksoy (2016), in a series of columns, reveals a highly detailed list of reasons behind the brain drain and the clear impact of increasing concern for the future of the country. This series of columns are based on interviews with current PhD students, scholars, and other people who have decided to leave the country or cannot return to Turkey because of some conflicts they encountered with the government in their past.

This negative environment has been worsened due to some other developments.[9] For example, it is reported that, talking to *Sunday's Zaman*, a high-level administrator said that more than 120 top-level scientists and administrators were fired from TÜBİTAK because they were disliked by the political powers, and most of them are looking for job opportunities abroad and are not planning to return for as long as the present political turmoil continues in Turkey (FactsonTurkey 2014).

Notes

1. Available at http://www.nsf.gov/statistics/2016/nsf16300/digest/nsf16300.pdf
2. See Yilmaz Şener and Elitok (2018) for a comprehensive comparative discussion on skilled Turkish return migration from Germany and the US.

3. For a comprehensive work on migration into Turkey, see Toksöz et al. (2012), and see Toksöz and Ünlütürk Ulutaş (2012) for its gender dimension.
4. Aydın (2012), on the other hand, provides an excellent discussion on the emigration of highly qualified Turks from Germany to Turkey.
5. This literature includes but not limited to Gülmez (1974), Dorsay (1977), Gökdere (1978), Erkal (1980), Gürgün (1980), Göker (1982), Şahinöz (1982), Kurtulmuş (1992), Gençler and Çolak (2002), Kaya (2002, 2009), Erdoğan (2003), Tuncel (2003), Fişek (2009), Çulpan (2005), Cansız (2006), Şimşek (2006), Babataş (2007), Gökbayrak (2004, 2006, 2008), İmeci (2009), Çengel (2009), Gündoğdu (2009), Özdemir (2009), Barışık and Çetinbaş (2009), Tansel and Güngör (2009), Sağırlı (2006), Sağbaş (2009), Güngör (2010), Bakırtaş and Kandemir (2010), Yıldırım (2010), Alvan (2012), and Özkan (2012).
6. Although it is not based on a questionnaire, İmeci (2009) can be considered in this group. İmeci (2009) investigates the return decision of 50 Ph.D. holders from Syracuse University in the US. Most of these scholars are from either electronic engineering or computer science programs. İmeci found that only 13 of them preferred to stay abroad. He notes that only nine of those students hold a government scholarship. That is, although a small ratio of students had compulsory service 74 percent of students returned to Turkey, and he interprets that as a very positive issue. İmeci (2009) notes that persons leave their countries due to more academic opportunities, greater employment opportunities, and better work atmosphere, as well as alienation from their home country for political atmosphere.
7. This is a law enacted by the Ministry of National Education in 1929 to form a government-sponsored study-abroad program to address the growing demand for qualified instructors at higher education (Ecnebi Memleketlere Gönderilecek Talebe Hakkında Kanun [Law Regarding Students to be Sent to Foreign Countries]).
8. The data used in Köser-Akçapar (2006) is based on a two-year research in different cities in the US and in Turkey. It is a part of the project supported by the Migration Research Program at Koç University (MiReKoc) in İstanbul and the Turkish Foundation of Social and Economic Studies (TESEV). Köser-Akçapar (2006) collects data from 'all available secondary data, onsite observation and inquiry, and semi-structured and in-depth interviews with (a) graduate students currently studying in different cities in the USA, (b) former students who have finished their studies and started working in the USA as young professionals between 25 and 45 years old, (c) those who came to the US 20 or 30 years ago and decided to stay for a number of reasons, (d) representatives of Turkish Students' Associations; (e) educational attaches and other government officials at the Turkish Embassy

in Washington, DC, at the Turkish General Consulate in New York City; and also with the General Consul of Houston and the General Consul of Los Angeles' (p. 6).
9. It is also argued that academic freedom has been under attack by the government authorities or there are various pressures/conflicts between scholars and authorities at university/institution level. For instance, during the proofread of this report, the newspapers in Turkey released news that reports the conflict between a Harvard graduate faculty and his university. http://t24.com.tr/haber/artuklu-universitesi-harvard-mezunu-akademisyenisten-cikardi-buldugu-prehistorik-taslari-gomme-karari-aldi,345976. A few days later in another news, although a project prepared by two high school students does not pass the initial qualification contest, it becomes the winner of the worldwide contest! http://www.haberturk.com/gundem/haber/1258992-tubitak-begenmedi-onlar-dunya-sampiyonu-oldu

REFERENCES

Abadan Unat, N. (2002). *Bitmeyen Göç: Konuk İşçilikten Ulus-Ötesi Yurttaşlığa.* İstanbul: İstanbul Bilgi Üniversitesi Yayınları.
Abakay, A. (1988). *Politik Göçmenler.* İstanbul: Amaç Yayıncılık.
Akman, V. (2014). Factors Influencing International Student Migration: A Survey and Evaluation of Turkey's Case. *Interdisciplinary Journal of Contemporary Research in Business,* 5(11), 390–415.
Aksoy, M. (2016). Neden gidiyorlar? Neden dönmüyorlar? www.haberdar.com; http://www.haberdar.com/yasam/neden-gidiyorlar-neden-donmuyorlar-h26170.html?mnst=9457. Accessed 11 July 2016.
Akyol, R. A. (2014). Reversing Turkey's brain drain. http://www.al-monitor.com/pulse/originals/2014/03/turkey-brain-drain-education-research.html#. Accessed 11 July 2016.
Altaş, D., Sağırlı, M., & Giray, S. (2006) Yurtdışında Çalışıp Türkiye'ye Dönen Akademisyenlerin Eğitim Durumları, Gidiş ve. Dönüş Sebepları Arasındaki İlişki Yapısının Loglineer Modeller ile İncelenmesi. *Marmara Üniversitesi İİBF Dergisi,* 21(1), 401–421.
Alvan, A. (2012). *Brain Drain: A Critical Analysis for Turkish Higher Education.* Gediz Üniversitesi Yayınları 7, Gediz University, İzmir, Turkey.
Atabek, E. (1971, February 3). Amerika'da Türk Hekimleri. *Cumhuriyet Gazetesi.*
Atılgan, D. (1986). Beyin Göçü. *TKDB,* 35(3), 27–36.
Aydın, Y. (2012). Emigration of Highly Qualified Turks: A Critical Review of the Societal Discourses and Social Scientific Research. In S. Paçacı Elitok & T. Straubhaar (Eds.), *Turkey, Migration and the EU: Potentials, Challenges and Opportunities.* Hamburg: Hamburg University Press.

Babataş, G. (2007). Beyin Göçü ve Türkiye'nin Sosyo-Ekonomik Yapısının Beyin Göçüne Etkisi. *Marmara Üniversitesi Sosyal Bilimler Enstitüsü Dergisi*, 7(28), 263–266.
Bakırtaş, T., & Kandemir, O. (2010). Gelişmekte Olan Ülkeler ve Beyin Göçü: Türkiye Örneği. *Kastamonu Eğitim Dergisi*, 18(3), 961–974.
Barışık, S., & Çetinbaş, H. (2009). Beyin Göçünde Yükseköğrenim, Ar-Ge Faaliyetleri, Çokuluslu Şirketlerin Önemi, *Eğitime Bakış*, 13, 38–46.
Başaran, F. (1972a). *Brain-Drain Problem in Turkey.* Ankara Üniversitesi Yayınevi. http://www.politics.ankara.edu.tr/dergi/tybook/13/Fatma_Basaran.pdf
Başaran, F. (1972b). Türkiye'de Beyin Göçü Sorunu. *Araştırma Ankara Üniversitesi Dil ve Tarih-Coğrafya Fakültesi Felsefe Bölümü Dergisi*, 10, 83–132.
Bilgili, Ö. (2012). *Turkey's Multifarious Attitude towards Migration and its Migrants.* Migration Policy Center Analytical and Synthetic Note 2012/01.
Bilgili, Ö., & Siegel, M. (2011). *Understanding the changing role of the Turkish diaspora.* UNU-MERIT Working Paper Series, 2011-039.
Bilgili, Ö., & Siegel, M. (2013). Policy perspectives of Turkey towards return migration: From permissive indifference to selective difference. *Migration Letters*, 11(2), 218–228.
Cansız, A. (2006). Son Yıllarda Beyin Göçünün Türk Yüksek Öğretimi Üzerindeki Etkileri, TMMOB Elektrik-Elektronik Bilgisayar Mühendislikleri Eğitimi 3. Ulusal, İstanbul. http://www.emo.org.tr/ekler/0353558f3ae8b91_ek.pdf. Accessed 11 July 2016.
Cumhuriyet Gazetesi. (2002, May 7). Beyaz Yaka Krizi. *Cumhuriyet Gazetesi.*
Çelik, S. (2012). Turkey's Ministry of National Education study-abroad program: Is the MoNE making the most of its investment? *The Qualitative Report*, 17(40), 31–40.
Çengel, Y. (2009). Beyin Gücü ve Beyin Göçü: Madalyonun Diğer Yüzü. *Eğitime Bakış*, 13, 7–13.
Çizmeci, Y. (1988). *Politik Göçmenler.* In A. Abakay (Ed.), ,*Politik Göçmenler.* İstanbul: Amaç Yayıncılık.
Çulpan, R. (2005). Turning Brain Drain into Brain Gain: Some Suggestions for Turkey. The TASSA Conference, Washington, DC.
Dedeoğlu, S. (2014). *Migrants, Work and Social Integration: Women's Labour in the Turkish Ethnic Economy in London.* Palgrave Macmillan.
Dedeoğlu, S., & Gökmen, Ç. E. (2011). *Göç ve Sosyal Dışlanma.* Ankara: Eflatun Yayınları.
Dirican, R., Taylor, C. E., & Deushle, K. W. (1968). *Health Manpower Planning in Turkey: An International Research Case Study.* Johns Hopkins University Press: Baltimore, MD.
Dorsay, Ş. (1977). Beyin Göçü. *İşveren*, 15(11), 21–23.
DPT. (1963). *Kalkınma Planı (Birinci Beş Yıl) (1963–1967).* Ankara: Başbakanlık Devlet Matbaası.

DPT. (1967). *Kalkınma Planı, İkinci Beş Yıl 1968–1972*. Ankara: Başbakanlık Devlet Matbaası.
DPT. (1972). *Yeni Strateji ve Kalkınma Planı, Üçüncü Beş Yıl 1973–1977*. Ankara: Başbakanlık Basımevi.
DPT. (1979). *Dördüncü Beş Yıllık Kalkınma Planı 1979–1983*. Ankara: Başbakanlık Basımevi.
DPT. (1981). *1981 İcra Planı, Dördüncü Beş Yıllık Kalkınma Planı 1979–1983*. Ankara: Başbakanlık Basımevi.
DPT. (1985a). *Beşinci Beş Yıllık Kalkınma Planı 1985–1989*. Ankara: Başbakanlık Basımevi.
DPT. (1985b). *V. Beş Yıllık Kalkınma Planı Öncesinde Gelişmeler 1972–1983, (Ekonomik ve Sosyal Gelişmeler)*. Ankara: DPT Yayınları.
DPT. (1989). *Altıncı Beş Yıllık Kalkınma Planı 1990–1994*. Ankara: Başbakanlık Basımevi.
DPT. (1995). *Yedinci Beş Yıllık Kalkınma Planı 1996–2000*. Ankara: Başbakanlık Basımevi.
DPT. (2000a). *Sekizinci Beş Yıllık Kalkınma Planı 2001–2005, Özel İhtisas Komisyonu Raporu*. Ankara: Başbakanlık Basımevi.
DPT. (2000b). *Uzun Vadeli Strateji ve Sekizinci Beş Yıllık Kalkınma Planı 2001–2005*. Ankara: DPT Yayınları.
DPT. (2006). *Dokuzuncu Kalkınma Planı 2007–2013.*Ankara: DPT Yayınları.
DPT. (2013). *Onuncu Kalkınma Planı 2014–2018*. Ankara: DPT Yayınları.
Elveren, A. Y., & Toksöz, G. (2017). Why Don't Highly Skilled Women Want to Return? Turkey's Brain Drain from a Gender Perspective. MPRA No: 80290, 2017. https://mpra.ub.uni-muenchen.de/80290/
Erdoğan, İ. (2003). Brain Drain and Turkey. *Kuram ve Uygulamada Eğitim Bilimleri/Educational Sciences: Theory & Practice*, 3(1), 96–100.
Erkal, M. (1980). Beyin Göçü. *Sosyoloji Konferansları Dergisi*, 18, 73–80.
Esen, E. (2014). *Going and Coming: Why U.S.-Educated Turkish PhD Holders Stay in the U.S. or Return to Turkey?* Unpublished PhD dissertation, Department of Educational Leadership and Policy Studies, The University of Kansas, Lawrence, KS, USA.
FactsonTurkey. (2014). Brain drain in Turkey due to political polarization, discrimination. http://factsonturkey.org/13106/brain-drain-in-turkey-due-to-political-polarization-discrimination/. Accessed 11 July 2016.
Fişek, G. (2009). Bir Danışıklı Döğüş: Beyin Göçü. *Eğitime Bakış, Sayı*, 13, 35–37.
Gençler, A., & Çolak, A. (2002). *Türkiye'den Yurtdışına Beyin Göçü: Ekonomik ve Sosyal Etkileri*. I. Ulusal Bilgi, Ekonomi ve Yönetim Kongresi Bildiriler Kitabı, İzmit, Turkey.
Gökbayrak, Ş. (2004). Beyin Göçünün Göç Veren Ülkeler Üzerindeki Etkilerine ve Politik Yaklaşımlarına İlişkin Güncel Tartışmalar. *Çalışma Ortamı*, 75, 19–23.

Gökbayrak, Ş. (2006). *Gelişmekte Olan Ülkelerden Gelişmiş Ülkelere Nitelikli İşgücü Göçü ve Politikalar- Türk Mühendislerinin "Beyin Göçü" Üzerine Bir İnceleme*. Unpublished PhD Dissertation, Department of Labor Economics and Industrial Relations, Ankara University, Ankara, Turkey.

Gökbayrak, Ş. (2008). Uluslararası Göç ve Kalkınma Tartışmaları: Beyin Göçü Üzerine Bir İnceleme. *Ankara Üniversitesi SBF Dergisi*, 63(3), 65–82.

Gökbayrak, Ş. (2009). Skilled Labour Migration and Positive Externality: The Case of Turkish Engineers Working Abroad. *International Migration*, 50(S1), 132–150.

Gökdere, M. (1978). Beyin Göçü Üzerine Bazı Düşünceler. İşveren, 16(5), 23–28.

Göker, A. (1982). Beyin Göçü. *Bilim ve Sanat*, 17, 7–10.

Güler, H. (2009). Tersine Beyin Göçü Artıyor [Reversal brain drain is increasing]. http://www.haberturk.com/yasam/haber/122194-tersine-beyin-gocu-artiyor. Accessed 11 July 2016.

Gülmez, M. (1974). Beyin İhracına Dönüşen Beyin Göçü. *Amme İdaresi Dergisi*, 7, 61–79.

Gündoğdu, A. (2009). Beyin Göçü ile Kaybettiğimiz Bilgi. *Eğitime Bakış*, 5(13), 1–2.

Güngör, N. D. (2003). *Brain drain from Turkey: An empirical investigation of the determinants of skilled migration and student non-return*. Unpublished PhD dissertations, Department of Economics, Middle East Technical University, Ankara, Turkey.

Güngör, N. D. (2010). Eğitim, Küreselleşme ve Beyin Göçü. *İz Atılım Üniversitesi Dergisi*, 9, 33–36.

Güngör, N. D. (2014). The Gender Dimension of Skilled Migration and Return Intentions. Invited Presentation, Harnessing Knowledge on the Migration of Highly Skilled Women: An Expert Group Meeting, International Organization for Migration and OECD Development Center, Geneva, Switzerland, 3–4 April 2014.

Güngör, N. D., & Tansel, A. (2008a). Brain Drain from Turkey: An Investigation of Students' Return Intention. *Applied Economics*, 40, 3069–3087.

Güngör, N. D., & Tansel, A. (2008b). Brain Drain From Turkey: The Case of Professionals Abroad. *International Journal of Manpower*, 29, 323–347.

Güngör, N. D., & Tansel, A. (2014). Brain Drain From Turkey: Return Intentions of Skilled Migrants. *International Migration*, 52(5), 208–226.

Gürgün, M. (1980). *Üstün Beyin Gücü Eğitimi ve Fen Lisesi Tercübesinin Sosyoekonomik Bakımdan Değerlendirilmesi*. Ankara: DPT.

İçduygu A. (2009). *International Migration and Human Development in Turkey*. MPRA Paper No. 19235, posted 12 December 2009. http://mpra.ub.uni-muenchen.de/19235/

İmeci, Ş. T. (2009). Beyin Göçü ve Türkiye. http://www.emo.org.tr/ekler/cd92dbca243f63e_ek.pdf. Accessed 11 July 2016.
Işığıçok, Ö. (2002). Türkiye'de yaşanan son ekonomik krizlerin sosyo-ekonomik sonuçları: Kriz işsizliği ve beyin göçü. *İş Güç Endüstri İlişkileri ve İnsan Kaynakları Dergisi*, 4(2).
Kamiloğlu, C. (2016). ABD'ye Yerleşmek İsteyen Türkler'in Sayısında Patlama. *Amerika'nın Sesi*. http://www.amerikaninsesi.com/a/abd-ye-yerlesmek-isteyen-turklerin-sayisinda-patlama/3339132.html. Accessed 11 July 2016.
Katseli, L. T., Lucas, R. E. B., & Xenogiani, T. (2006). *Effects of Migration on Sending Countries: What Do We Know?* OECD Development Centre, WP No. 250.
Kaya, M. (2002). *Beyin göçü/erozyonu*. Technology Research Centre Report, Osmangazi University, Eskişehir, Turkey.
Kaya, M. (2009). Beyin Göçü/Entellektüel Sermaye Erozyonu Bilgi Çağının Gönüllü Göçerleri: Beyin Gurbetçileri. *Eğitime Bakış*, 13, 14–29.
Kınıklıoğlu, S. (2014). Turkey's secular brain drain. http://www.al-monitor.com/pulse/politics/2014/11/turkey-secular-immigration-exodus.html. Accessed 11 July 2016.
Kösemen, C. (1968). *A Report presented to Education and World Affairs concerning Emigration of Engineers and Architects from Turkey to foreign countries for the project*. International Migration of Talent.
Köser-Akçapar, Ş. (2006). Do Brains really going down the Drain? Highly skilled Turkish Migrants in the USA and the «Brain Drain» Debate in Turkey. *Revue européenne des migrations internationales*, 22(3), 79–107.
Köser-Akçapar, Ş. (2009). Turkish Highly Skilled Migration to the United States: New Findings and Policy Recommendations. In A. İçduygu & K. Kirişci (Eds.), *Land of Diverse Migrations, Challenges of Emigration and Immigration in Turkey*. İstanbul: İstanbul Bilgi Üniversitesi Yayınları.
Kurtsal, Y. (2011). Türkiye'de Tersine Beyin Göçü Geçmişe ve Günümüze Bakış. *İstihdamda 3 İşgücü, İşveren, İşkur*, 3, 67–69.
Kurtulmuş, N. (1992). *Gelişmekte Olan Ülkeler Açısından Stratejik İnsan Sermayesi Kaybı: Beyin Göçü*. Sosyal Siyaset Konferansları Yayın No. 3662. İstanbul: İstanbul Üniversitesi.
Kurtuluş, B. (1999). *Amerika Birleşik Devletleri'ne Türk Beyin Göçü*. İstanbul: Alfa Basım Yayım Dağıtım.
Louscher, D. J., & Cook, A. H. (2001). *A Study on the US National Economic Impact from Turkish Students Attending US Educational Institutions*. Turkish-American Business Forum.
Mollahaliloğlu, S., Çulha, Ü. A., Kosdak, M., & Öncül, H. G. (2014). The Migration Preferences of Newly Graduated Physcians in Turkey. *Medical Journal of Islamic World Academy of Sciences*, 22(2), 69–75.

Oğuzkan, T. (1971). *Yurt Dışında Çalışan Doktoralı Türkler: Türkiye'den Başka Ülkelere Yüksek Seviyede Eleman Göçü Üzerinde Bir Araştırma*. Middle East Technical University, Ankara, Turkey.
Oğuzkan, T. (1975). The Turkish brain drain: Migration of tendencies among doctoral level manpower. In R. E. Krane (Ed.), *Manpower Mobility Across Cultural Boundaries: Social, Economic and Legal Aspects, The Case of Turkey and West Germany*. Leiden, Netherlands: E.J. Brill.
Oğuzkan, T. (1976). The Scope and Nature of Turkish Brain Drain. In N. Abadan-Unat (Ed.), *Turkish Workers in Europe, 1960–1975: A Socio-Economic Reappraisal* (pp. 74–103). Leiden, Netherlands: E.J. Brill.
Özdemir, İ. (2009). Beyin Göçünü Yeniden Düşünmek. *Eğitime Bakış*, 13, 3–6.
Özkan, M. (2012). Bilim politikası var mı ki geri dönelim! [Do you have policy on science that you are calling us?]. http://www.fp7.org.tr/home.do?ot=5&rt=&sid=0&pid=0&cid=24147. Accessed 11 July 2016.
Öztürk, M. (2001). *So far from home: Turkish Student Motivation and Access to Overseas Higher Education in the United States*. Unpublished PhD Dissertation, Faculty of Graduate School, University of Southern California, Los Angeles, CA, USA.
Paçacı Elitok, S., & Straubhaar, T. (Eds.). (2012). *Turkey, Migration and the EU: Potentials, Challenges and Opportunities*. Hamburg: Hamburg University Press.
Papila, N. (2015). New Models of Brain Circulation Half-time in the US and Half-time in Turkey—Is this a real possibility? http://www.tassausa.org/Newsroom/item/2124/Crossing-the-Bridge---March-2015. Accessed 11 July 2016.
Pazarcık, S. F. (2010). *Beyin Göçü Olgusu ve Amerika Birleşik Devletleri Üniversitelerinde Çalışan Türk Sosyal Bilimciler Üzerine Bir Çalışma*. Unpublished Master's Thesis, Çanakkale Onsekiz Mart Üniversitesi, Çanakkale, Turkey.
Sağbaş, S. M. (2009). *Beyin Göçünün Ekonomik ve Sosyal Etkileri: Türkiye Örneği*. Unpublished Master's Thesis, Department of Labor Economics and Industrial Relations, Marmara University, Istanbul, Turkey.
Sağırlı, M. (2006). *Eğitim ve İnsan Kaynağı Yönünden Türk Beyin Göçü: Geri Dönen Türk Akademisyenler Üzerine Alan Araştırması*. Unpublished PhD Dissertation, Department of Labor Economics and Industrial Relations, İstanbul University, İstanbul, Turkey.
Şahinöz, A. (1982). Beyin Göçü ve Türkiye. *Amme İdaresi Dergisi*, 15(4), 49–62.
Şimşek, M. (2006). *Beşeri Sermaye ve Beyin Göçü Kapsamında Türkiye: Karşılaştırmalı Bir Analiz*. Bursa, Turkey: Ekin Kitabevi.
Sunata, U. (2002). *Not A "Flight" from Home but Potential Brain Drain*. Unpublished Master's Thesis, Middle East Technical University, Ankara, Turkey.

Tansel, A., & Güngör, N. D. (2002). 'Brain Drain' from Turkey: Survey Evidence of Student Non-Return. http://ssrn.com/abstract=441160
Tansel, A., & Güngör, N. D. (2003). Brain Drain' from Turkey: Survey Evidence of Student Non-Return. *Career Development International*, 8(2), 52–69.
Tansel, A., & Güngör, N. D. (2004). *Türkiye'den Yurtdışı'na Beyin Göçü: Ampirik Bir Uygulama*. Economic Research Center Working Paper 04/02, METU, Ankara.
Tansel, A., & Güngör, N. D. (2009). Türkiye'den Beyin Göçü: Bir Anket Çalışması Üzerine Düşünceler. *Eğitime Bakış*, 13, 30–34.
Tanyıldız, Z. E., Arslanhan, S., & Kurtsal, Y. (2011). *Tersine Beyin Göçü Politikalarının Sanayi Politikalarına Entegrasyonu: Ülke Örnekleri ve Türkiye İçin Dersler*. TEPAV Politika Notu, Ankara, Turkey.
Tezcan, M. (1971). Beyin Göçü ve Türkiye. *Amme İdaresi Dergisi*, 4(3), 44–71.
Toksöz, G. (2006). *Uluslararası Emek Göçü*. İstanbul: İstanbul Bilgi Üniversitesi Yayınları.
Toksöz, G., Erdoğdu, S., & Kaşka, S. (2012). *Irregular Labour Migration in Turkeyand Situations of Migrant Workers in the Labour Market*. Sweden: International Organization of Migration.
Toksöz, G., & Ünlütürk Ulutaş, Ç. (2012). Is Migration Feminized? A Gender- and Ethnicity-Based Review of the Literature on Irregular Migration to Turkey. In S. Paçacı Elitok & T. Straubhaar (Eds.), *Turkey, Migration and the EU: Potentials, Challenges and Opportunities*. Hamburg: Hamburg University Press.
TÜBİTAK. (2010). *2011–2016 Bilim ve Teknoloji İnsan Kaynağı Stratejisi ve Eylem Planı*. Ankara.
TÜBİTAK. (2015). *TÜBİTAK'ın "Tersine Beyin Göçü Programı" Başarıya Ulaştı*. https://www.tubitak.gov.tr/tr/haber/tubitakin-tersine-beyin-gocu-programi-basariya-ulasti. Accessed 11 July 2016.
Tuncel, B. N. (2003). *Gelişmekte olan Ülkelerden Gelişmiş Ülkelere Beyin Göçü: Türkiye'de Bilgisayar Sektörü Örneği*. Unpublished Master's Thesis, Ankara University, Ankara, Turkey.
Turkish Chambers of Engineers and Architects. (1972). *Report*. Ankara.
Ünver, C. (2010). *Tersine Beyin Göçü Başladı (Reverse Brain Drain Has Begun)*. Uluslararası İlişkiler ve Stratejik Analizler Merkezi, Turkey.
Uysal, Ş. (1972). *Yurdışında Yetişen İhtisas Gücü Raporu*. TÜBİTAK Bilim Adamı Yetiştirme Grubu Proje No. Bayg-E-22.
Vatansever Deviren, N., & Daşkıran, F. (2014). Yurt Dışında Eğitim Görüp Geri Dönen Öğretim Elemanlarının Beyin Göçüne Bakışı: Muğla Sıtkı Koçman Üniversitesi Örneği. *Dumlupınar Üniversitesi Sosyal Bilimler Dergisi*, 41, 1–10.
Yavuzer, H. (2000). Eğitim Durumu ve Beyin Göçü Bakımından Amerika'daki Türkler. *Dergi Karadeniz*, 8, 89–107.
Yiğit, K. (2011). İİBK'dan İŞKUR'a Yurtdışına İşçi Göçü. *İstihdamda 3 İ İşgücü, İşveren, İşkur*, 3, 44–47.

Yıldırım, T. (2010). Uluslararası Düzeyde Sağlık Çalışanlarının Göçünü Yönetme Politikaları: Genel Bir Bakış ve Türkiye İçin Bir Durum Değerlendirmesi. *Amme İdaresi Dergisi*, 43, 31–65.

Yilmaz Şener, M., & Pacaci Elitok, S. (2018). Getting Adapted? A Comparative Study of Qualified Turkish Return Migrants from Germany and the USA. In M. Caselli & G. Gilardoni (Eds.), *Globalization, Supranational Dynamics and Local Experiences. Europe in a Global Context*. Palgrave Macmillan.

Yurtdışı Türkler ve Akraba Topluluklar Başkanlığı. (2018). http://www.ytb.gov.tr/

Zweig, D., & Changgui, C. (1995). *China's Brain Drain to the United States: Views of Overseas Chinese Students and Scholars in the 1990s*. China Research Monograph, Institute of East Asian Studies. Berkeley, CA: University of California.

CHAPTER 4

A Brain Drain Survey

Abstract This chapter analyzes the return intentions of undergraduate and graduate students and professionals by using an original survey data. The chapter employs different statistical techniques such as Chi-square test, correspondence analysis, basic correlation tests, and ordered probit model to show the determinants of migration and (non)return decision regarding study or work field, previous work experience, and so on. The chapter investigates the gender differentiation in return intentions.

Keywords Brain drain • Gender • Chi-square test • Correspondence analysis • Regression

The data of this study is obtained from an Internet survey based on referral and snowball sampling methods. We acknowledge that referral sampling may have some biases in responses. Considering a relatively small number of respondents, we also acknowledge that our sample is certainly not a representative of the entire population of Turkish students/professionals residing abroad. However, despite these basic and common shortcomings, the study provides some valuable and recent information on the phenomenon of the brain drain as it reaches out as many as 200 students/professionals from various fields to collect detailed information about their intention to return to Turkey. The detailed descriptive statistics and the results of Chi-square tests are presented in Sect. 4.1. The results of statistical analyses are provided in Sect. 4.2.

4.1 Descriptive Statistics

The below presents detailed descriptive statistics of the students and professionals and the responses to the questions that aim to measure students' and professionals' intention (not) to return. The data is collected via a questionnaire[1] of 46 questions for professionals in English and a questionnaire of 53 questions for students both in Turkish and in English that were online and accessible to everyone from the late 2015 to the early 2016 for about five months. The statistics are based on 116 students and 84 professionals who replied all questions in the surveys. When possible, statistics are provided by gender because this study aims at contributing to the literature by adding more information on the gender aspects of the brain drain, a topic that has not received enough attention.

Table 4.1 shows the age and gender percentages of students and professionals. There are two main characteristics of the sample. First, not just students but also professionals are relatively young. Eighty-two percent of students are younger than 30, and 77.4 percent of professional are younger than 40 years old.[2] Second, there is relatively equal distribution in terms of gender. Totally 41.4 (58.6) percent of students and 33.3 (66.7) percent of professionals are female (male). Overall, 38 percent of all participants are female, and 62 percent are male.

One-fifth of students and one-third of professionals were born in Ankara, followed by İstanbul. In case of the professionals, as high as 63 percent of professionals are from Ankara, İstanbul, or İzmir. This is not an unexpected result as one can see the similar results in studies reviewed in Chap. 3. However, while about 40 percent of students were born in Ankara, İstanbul, or İzmir, the rest (about 60 percent) is from different

Table 4.1 Age and gender (%)

Age	Students			Professionals		
	Male	Female	Total	Male	Female	Total
<26	27.9	29.2	28.5	1.8	3.6	2.4
26–30	52.9	54.2	53.5	23.2	3.6	16.7
31–40	19.1	14.6	17.1	51.8	71.4	58.3
>40	0	2.1	0.9	23.2	21.4	22.6
n	68	48	116	56	28	84
Percentage	58.6	41.4	100	66.7	33.3	100

provinces across the country. This is likely to be a result of extensive scholarships by the Ministry of Education.

As expected, a large part of both students (96.3 percent in total) and professionals (91.6 percent in total) resides in the US, followed by the UK and Canada.

Only as low as 8.3 percent of professionals' highest degree is a bachelor's. Around 61.9 percent of them have Ph.D. and 26.2 percent of them have master's degree. Overall 41.7 percent of professionals are academic, while 58.3 percent has nonacademic positions. Among academics there are 2 professors, 7 associate professors, 12 assistant professors, 11 instructors/lecturers, and 25 research/teaching assistants.

Table 4.2 shows that while only 17.9 percent of professionals received their highest degree from an institution in Turkey, as high as 73.8 percent of them earned their highest degree in the US. This is an important statistic, whose relationship with the return intention should be analyzed. On the other hand, while 64.7 percent of students earned their highest degree from Turkey, 26.7 percent received from the US.

A similar picture regarding the birth city of students can be seen in the distribution of bachelor's degree institution as well. While about 61 percent of the professionals earned their bachelor's degree from Middle East Technical University, Boğaziçi University, Bilkent University (top three universities commonly reported in other brain drain surveys), and foreign universities, the bachelor's institutions for students show a much wider spread across the country.

Table 4.2 Country of highest degree

	Students			Professionals		
	Freq.	Percent		Freq.	Percent	
Turkey	75	64.7	United S	62	73.8	
US	31	26.7	Turkey	15	17.9	
UK	4	3.5	UK	3	3.6	
Australia	1	0.9	Canada	1	1.2	
Canada	1	0.9	Germany	1	1.2	
France	1	0.9	Netherlands	1	1.2	
Germany	1	0.9	Switzerland	1	1.2	
Poland	1	0.9				
Switzerland	1	0.9				
Total	116	100	Total	84	100	

There is a similar distribution of students' and professionals' majors and their current study field. In both groups the majority (37.6 percent of students and 52.4 percent of professionals) is engineering and technical sciences. However, it is a favorable aspect of this study that it does not focus on either natural science or social sciences or a certain occupational category such as doctors or engineers. In fact, while 56.4 percent of students are in natural sciences, 43.6 percent are in social sciences. The same ratios are 60.7 percent and 39.3 percent for professionals, respectively. This relatively balanced distribution of fields allows us to investigate possible differentiation between these two broad categories of fields.

About 69.8 percent of students and 72.6 percent of professionals had some previous experience abroad (any combination of study, work, and travel). These ratios, particularly for students, are relatively high. This is an initial sign that either students are from relatively well-doing families so that they had an opportunity to go abroad and/or they were already open to the idea of going abroad. In the case of the students, 86.5 percent of these experienced a period of longer than a month. The same ratio was 88.6 for the professionals.

One of the major questions of the survey is 'What were your main reasons for going to the country you are currently residing in?' The most popular reasons for going abroad for students are 'prestige and advantages of study abroad' (75.9 percent), 'insufficient facilities, equipment for research in Turkey' (52.6 percent), 'learn language, improve language skills' (46.6 percent), and 'get away from political environment in Turkey' (44.0 percent). In the case of professionals, the answers are relatively similar, with one crucial difference: 'prestige and advantages of study abroad' (60.7 percent), 'insufficient facilities, equipment for research in Turkey' (40.5 percent), 'get away from political environment in Turkey' (40.5 percent), and finally 'lifestyle preference' (35.7 percent). That is, for professionals 'lifestyle preference' is also a major reason for going/staying abroad. Perhaps, a second difference between students and professionals in terms of their reasons for going abroad can be their different emphasize on the 'get away from political environment in Turkey'.

To better understand this difference one can check Tables 4.3 and 4.4, where the participants answer the most important reason for going abroad. Both for students and for professionals the most important reason is reported to be 'prestige and advantages of study abroad', and 'get away from political environment in Turkey' is not one of popular answers in

Table 4.3 The most important reason for going abroad (%)

	Students			Professionals		
	Male	Female	Total	Male	Female	Total
Learn language, improve language skills	8.8	6.3	7.8	3.6	3.6	3.6
Need change, experience new culture	2.9	4.2	3.5	3.6	10.7	6.0
Job requirement in Turkey	10.3	6.3	8.6	19.6	21.4	20.2
Could not find a job in Turkey	0.0	2.1	0.9	1.8	0.0	1.2
No program in specialization in Turkey	4.4	0.0	2.6	3.6	3.6	3.6
Insufficient facilities, equipment for research in Turkey	19.1	20.8	19.8	8.9	10.7	9.5
Prestige and advantages of study abroad	36.8	39.6	37.9	26.8	25.0	26.2
Lifestyle preference	11.8	6.3	9.5	14.3	7.1	11.9
To be with spouse/loved one	0.0	4.2	1.7	8.9	14.3	10.7
Provide better environment for children	0.0	2.1	0.9	1.8	0.0	1.2
Get away from political environment in Turkey	5.9	4.2	5.2	7.1	3.6	6.0
Other	0.0	4.2	1.7	0.0	0.0	0.0
n	*68*	*48*	*116*	*56*	*28*	*84*

Students: Chi2(11) = 12.678; professionals: Chi2(10) = 4.415

Table 4.4 The most important reason for going abroad (%) (students + professionals)

	Male	Female	Total
Learn language, improve language skills	6.5	5.3	6.0
Need change, experience new culture	3.2	6.6	4.5
Job requirement in Turkey	14.5	11.8	13.5
Could not find a job in Turkey	0.8	1.3	1.0
No program in specialization in Turkey	4.0	1.3	3.0
Insufficient facilities, equipment for research in Turkey	14.5	17.1	15.5
Prestige and advantages of study abroad	32.3	34.2	33.0
Lifestyle preference	12.9	6.6	10.5
To be with spouse/loved one	4.0	7.9	5.5
Provide better environment for children	0.8	1.3	1.0
Get away from political environment in Turkey	6.5	4.0	5.5
Other	0.0	2.6	1.0
n	*124*	*76*	*200*

Chi2(11) = 10.073

Table 4.3. This finding implies that answers should be evaluated carefully before jumping to the conclusion.

Table 4.4 shows that there is not statistical significant differentiation between male and female participants in terms of the most important reason for going abroad. It is also true in the case of students and professionals (Table 4.3).

There is an interesting differentiation between students and professionals in terms of family support for the decision to go abroad. There are two remarkable differences between the subgroups of the participants. First, while there is a highly balanced distribution of different support levels for the students, the support for professionals was substantially higher. While 50 percent of students' families were either 'very supportive' or 'supportive', the ratio was as high as 89.3 for the professionals. Second, while there is not a remarkable difference between genders for the professionals, female students seem to receive much higher support from their parents/family to go abroad. While 42.6 percent of male students' families were either 'very supportive' or 'supportive', the same ratio was 60.5 percent for female students.

Table 4.5 presents the reasons for return for those who reported their intention to return. Three most common reasons for students are 'missing family', 'achieving specific goals (e.g., work experience)', and 'complete

Table 4.5 Return reasons (%)

	Students			Professionals		
	Male	Female	Total	Male	Female	Total
Complete military service	1.8	0.0	1.1	0.0	0.0	0.0
Complete university service	41.1	63.6	49.4	3.6	10.0	5.3
Expiry of overseas job contract	12.5	0.0	7.9	3.6	10.0	5.3
Miss family	53.6	48.5	51.7	57.1	50.0	55.3
Children's education	19.6	12.1	16.9	10.7	10.0	10.5
Achieve specific goals (e.g., work experience)	55.4	45.5	51.7	39.3	10.0	31.6
Achieve savings goal	21.4	3.0	14.6	17.9	0.0	13.2
Achieved career goal	0.0	0.0	0.0	39.3	30.0	36.8
Job opportunity in Turkey	1.8	0.0	1.1	3.6	0.0	2.6
Retirement	0.0	0.0	0.0	42.9	40.0	42.1
Lack of safety in current environment	12.5	3.0	9.0	7.1	0.0	5.3
Other	0.0	0.0	0.0	10.7	10.0	10.5
n	56	33	89	28	10	38

university service' in order of importance. For professionals on the other hand, they are 'missing family', 'retirement', and 'achieving career goals'.

Those who reported their return intentions are also asked if they plan to return abroad again (i.e., after you return, do you plan to go abroad again?). Tables 4.6 and 4.7 demonstrate the answers for this question by gender. One remarkable issue here is a much higher percentage of 'yes' among students (31.8 percent) than among professionals (13.2 percent). Another significant remark presented in Table 4.6 is the difference between genders for both students and professionals. While 24.5 percent of male students reported that they plan to go to abroad again, the same ratio is as high as 43.8 percent for female students. Similarly, ratio of female students who plan to go abroad again is virtually twice higher than that for male students (10.7 percent vs. 20.0 percent).

Table 4.8 presents three information, marital status of the participant, the job status, and nationality of the spouse if the participant is married. Twenty-six percent of students are married, 79.3 percent of them have Turkish spouses, and 20.7 percent of the spouses have foreign nationality. Around 62.1 percent of spouses of the students have a full-time job.

Table 4.6 Plan to go abroad again and gender (%)

	Students			Professionals		
	Male	Female	Total	Male	Female	Total
Yes	24.5	43.8	31.8	10.7	20.0	13.2
Maybe	52.8	43.8	49.4	75.0	40.0	65.8
No	22.6	12.5	18.8	14.3	40.0	21.2
n	53	32	85	28	10	38

Students: Chi2(2) = 3.744; professionals: Chi2(2) = 4.169

Table 4.7 Plan to go abroad again and gender (%) (students + professionals)

	Male	Female	Total
Yes	19.8	38.1	26.0
Maybe	60.5	42.9	54.5
No	19.8	19.1	19.5
n	81	42	123

Chi2(2) = 5.163*; * refers to significance at the 10% level

Table 4.8 Marital status and the status of spouse

	Students		Professionals	
	Freq.	Percent	Freq.	Percent
Never married	78	70.9	22	26.5
Divorced/widowed/separated	3	2.7	5	6.0
Married	29	26.0	56	67.5
Spouse's nationality = Turkish	23	79.3	39	69.6
Spouse's nationality = foreign	6	20.7	13	23.2
Spouse's nationality = dual citizen	0	0.0	4	7.1
Spouse's employment = full time	18	62.1	41	73.2
Spouse's employment = part time	0	0.0	4	7.1
Spouse's employment = not employed	11	37.9	11	19.6
Total	110	100	83	100

Table 4.9 Effect of 2008–2009 economic crisis in the US on views about returning

	Students		Professionals	
	Freq.	Percent	Freq.	Percent
Increased my likelihood of returning	2	1.8	5	6.0
Decreased my likelihood of returning	4	3.6	5	6.0
Did not change my views	39	35.5	58	69.9
Not applicable	65	59.1	15	18.1
Total	110	100	83	100

Overall 67.5 percent of professionals are married, 69.6 percent of whose spouses are Turkish, 23.2 percent are foreign, and 7.1 percent are dual citizens. Totally 73.2 percent of spouses of professionals are full-time employed.

Table 4.9 clearly shows that there is not significant effect of the 2008–2009 economic crisis in the US on the return intentions of neither students nor professionals.

4.2 A Statistical Analysis

This section presents the results of statistical analysis. We utilize basic correlations test, correspondence analysis, and a regressions analysis to examine the determinants of the return intentions. One of the most significant

Table 4.10 Initial-current comparison, (%) students

	n	Initial		
		Return	Undecided	Stay
Return immediately after studies	30	41.0	12.8	0.0
Definitely return but not soon after studies	24	23.0	20.5	12.5
Return probable	37	23.0	48.7	25.0
Return unlikely	22	11.5	18.0	50.0
Definitely not return	3	1.6	0.0	12.5
Total	116			

results of this research is the positive relationship between the initial return decision and the current return decision. This confirms the findings of Gökbayrak (2009) and Güngör and Tansel (2008a, b, 2014). This significant relationship is presented in Tables 4.10, 4.11, and 4.12.

Table 4.12 shows that there exists positive relationship between the current return intentions and the initial return intentions, for all groups and genders according to three basic correlation tests.

Tables 4.13 and 4.14 both show that although there is positive correlation between 'stay duration abroad' and 'the initial non-return intention' for only students and females, that relationship, as expected, is more significant in the case of 'stay duration' and 'the current non-return intention' for all groups. This is a highly significant result that shows the impact of time spent abroad on non-return decisions, supporting Güngör and Tansel (2008a, b, 2014).

As expected there is not a significant correlation between age and initial return decision (Table 4.15). However, as presented in Table 4.16, except for professionals, for all other groups the higher age is positively associated with the current non-return decision. This is also in line with Güngör and Tansel (2008a, b, 2014). This is because the psychic cost of migration might be higher for older persons as argued by Güngör (2003).

Tables 4.17, 4.18, 4.19, and 4.20 show the relationship between the current return intentions and the country from which students/professional received their highest degree. Tables show that students or professionals who received their degree from a foreign university are more likely not to return. This finding also supports Güngör and Tansel (2008a, b, 2014). The same relationship does not exist in the case of initial return intentions (Tables 4.21, 4.22, 4.23, and 4.24).

Table 4.11 Initial-current comparison, (%) professionals

	n	Initial		
		Return	Undecided	Stay
Definitely return, plans	3	2.5	3.9	5.6
Definitely return, no plans	2	2.5	3.9	0.0
Return probable	33	57.5	23.1	22.2
Return unlikely	34	27.5	65.4	33.3
Definitely not return	12	10.0	3.9	38.9
Total	84			

Table 4.12 Initial return intention, current return intention

Sample	Pearson	Spearman	Kendall
All	0.370***	0.361***	0.320***
Students	0.456***	0.441***	0.397***
Professionals	0.257**	0.311***	0.287***
Male	0.376***	0.370***	0.325***
Female	0.368***	0.336***	0.304***

** and *** refer to significance at the 5% and 1% levels, respectively

Table 4.13 Initial non-return intention, stay duration

Sample	Pearson	Spearman	Kendall
All	0.166**	0.132*	0.114*
Students	0.245***	0.209**	0.183**
Professionals	0.050	0.020	0.014
Male	0.092	0.063	0.052
Female	0.293**	0.238**	0.211**

*, ** and *** refer to significance at the 10%, 5% and 1% levels, respectively

Table 4.14 Current non-return intention, stay duration

Sample	Pearson	Spearman	Kendall
All	0.339***	0.310***	0.256***
Students	0.394***	0.398***	0.331***
Professionals	0.283***	0.240**	0.204**
Male	0.297***	0.289***	0.236***
Female	0.408***	0.318***	0.270***

** and *** refer to significance at the 5% and 1% levels, respectively

Table 4.15 Initial non-return intention, age group

Sample	Pearson	Spearman	Kendall
All	0.074	0.430	0.037
Students	−0.057	−0.076	−0.067
Professionals	0.135	0.130	0.118
Male	0.071	0.043	0.037
Female	0.078	0.043	0.037

Table 4.16 Current non-return intention, age group

Sample	Pearson	Spearman	Kendall
All	0.412***	0.414***	0.352***
Students	0.236**	0.237**	0.206**
Professionals	0.104	0.089	0.080
Male	0.408***	0.416***	0.348***
Female	0.430***	0.421***	0.371***

** and *** refer to significance at the 5% and 1% levels, respectively

Table 4.17 Current return—highest degree country, (%) students

	n	Highest degree	
		Turkey	Foreign
Return immediately after studies	30	34.7	9.8
Definitely return but not soon after studies	24	22.7	17.1
Return probable	37	29.3	36.6
Return unlikely	22	10.7	34.2
Definitely not return	3	2.7	2.4
Total	116		

Chi2(4) = 14.909***; *** refers to significance at the 1% level

Table 4.18 Current return—highest degree country, (%) professionals

	n	Highest degree	
		Turkey	Foreign
Definitely return, plans	3	20.0	0.0
Definitely return, no plans	2	6.7	1.5
Return probable	33	53.3	36.2
Return unlikely	34	6.7	47.8
Definitely not return	12	13.3	14.5
Total	84		

Chi2(4) = 21.294***; *** refers to significance at the 1% level

Table 4.19 Current return—highest degree country (%) (students + professionals)

	Highest degree	
	Turkey	Foreign
Definitely return, plans	32.2	3.6
Definitely return, no plans	20.0	7.3
Return probable	33.3	36.4
Return unlikely	10.0	42.8
Definitely not return	4.4	10.0
n	90	110

Chi2(4) = 51.784** refers to significance at the 10% level; *** refers to significance at the 1% level

Table 4.20 Current non-return intention, highest degree

Sample	Pearson	Spearman	Kendall
All	0.402***	0.414***	0.354***
Students	0.240***	0.275***	0.231***
Professionals	0.239**	0.230**	0.207**
Male	0.471***	0.483***	0.409***
Female	0.295***	0.304***	0.263***

** and *** refer to significance at the 5% and 1% levels, respectively

Table 4.21 Initial return—highest degree country, (%) students

	n	Highest degree	
		Turkey	Foreign
Return	61	53.3	51.2
Undecided	39	36.0	29.3
Stay	16	10.7	19.5
Total	116		

Chi2(2) = 1.884

Table 4.22 Initial return—highest degree country, (%) professionals

	n	Highest degree	
		Turkey	Foreign
Return	40	46.7	47.8
Undecided	26	26.7	31.9
Stay	18	26.7	20.3
Total	84		

Chi2(2) = 0.346

Table 4.23 Initial return—highest degree country (%) (students + professionals)

	Highest degree	
	Turkey	Foreign
Return	52.2	49.1
Undecided	34.4	30.9
Stay	13.3	20.0
n	90	110

Chi2(2) = 1.581

Table 4.24 Initial non-return intention, highest degree country

Sample	Pearson	Spearman	Kendall
All	0.011	−0.009	−0.007
Students	0.055	0.037	0.034
Professionals	−0.174	−0.181*	−0.165*
Male	0.064	0.036	0.033
Female	−0.073	−0.083	−0.073

* refers to significance at the 10% level

Tables 4.25, 4.26, and 4.27 present the relationship between current return intention and bachelor major in two main categories: natural science and social science. Figures 4.1, 4.2, 4.3, 4.4, and 4.5 present the same relationship for four groups of majors. Accordingly, compared to natural science majors, social science (both students and professionals together) graduates are more likely to return (Table 4.27).

Figures 4.1, 4.2, 4.3, 4.4, and 4.5 show the correspondence analysis, in which ENG refers to 'engineering-technical-city plan architect', MNS refers to 'mathematics and natural sciences, medical health', EAS refers to 'economics and administrative sciences', EDUC refers to 'educational sciences', and other SOC refers to 'other social sciences, including art'. The other dimension is the current return intentions, where DRP refers to 'definitely return, plans', DRnP refers to 'definitely return, no plans', RP refers to 'return probable', RU refers to 'return unlikely', and NotR refers to 'definitely not return'. Accordingly, 'mathematics and natural sciences, medical health' majors are more likely not to return (Figs. 4.1 and 4.2), and 'engineering-technical-city plan architect' majors are more likely to

Table 4.25 Current return—bachelor degree discipline, (%) students

	n	Bachelor degree	
		Natural science	Social science
Return immediately after studies	24	15.8	34.1
Definitely return but not soon after studies	21	31.6	6.8
Return probable	35	29.8	40.9
Return unlikely	19	22.8	13.6
Definitely not return	2	0.0	4.6
Total	101		

Chi2(4) = 15.404***; *** refers to significance at the 1% level

Table 4.26 Current return—bachelor degree discipline, (%) professionals

	n	Bachelor degree	
		Natural science	Social science
Definitely return, plans	3	0.0	9.1
Definitely return, no plans	2	0.0	6.1
Return probable	33	49.0	24.2
Return unlikely	34	35.3	48.5
Definitely not return	12	15.7	12.1
Total	84		

Chi2(4) = 11.898**; ** refers to significance at the 5% level

Table 4.27 Current return—bachelor degree discipline (%) (students + professionals)

	Bachelor degree	
	Natural science	Social science
Definitely return, plans	8.3	23.4
Definitely return, no plans	16.7	6.5
Return probable	38.9	33.8
Return unlikely	28.7	28.6
Definitely not return	7.4	7.8
n	108	77

Chi2(4) = 11.042**; ** refers to significance at the 5% level

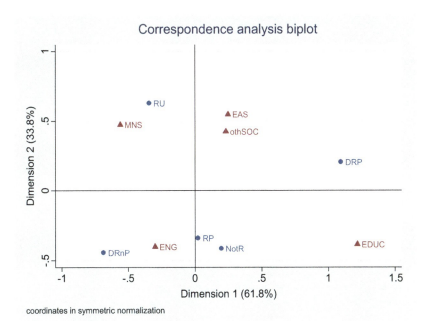

Fig. 4.1 Current return intention—bachelor degree (all sample)

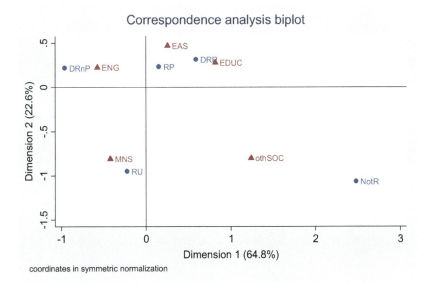

Fig. 4.2 Current return intention—bachelor degree (students)

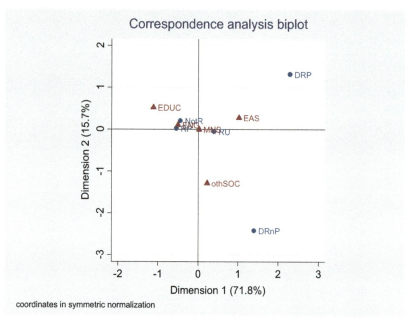

Fig. 4.3 Current return intention—bachelor degree (professionals)

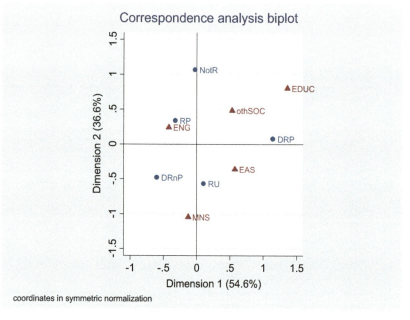

Fig. 4.4 Current return intention—bachelor degree (male)

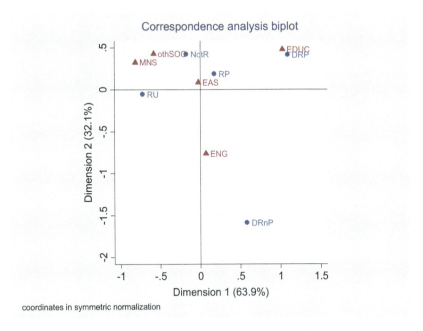

Fig. 4.5 Current return intention—bachelor degree (female)

return (Figs. 4.1 and 4.2), and there is no remarkable tendency/pattern for gender (Figs. 4.4 and 4.5).

Tables 4.28, 4.29, and 4.30 present the relationship between the initial return intention and bachelor majors in two main categories, natural science and social science. Figures 4.6, 4.7, 4.8, 4.9, and 4.10 present the same relationship for four groups of majors. Tables show no significant relationship between the variables. Figures 4.6, 4.7, 4.8, 4.9, and 4.10 demonstrate two tendencies. While 'economics and administrative science' majors have tendency to return to home country, 'mathematics and natural science, medical health' majors have tendency to stay.

Tables 4.31, 4.32, and 4.33 show the relationship between the current return decision and the field of study/work. Although there is a significant difference between social science and natural science, the tendency to stay or return is ambiguous. Figures 4.11, 4.12, 4.13, 4.14, and 4.15 show the results of correspondence analysis of the same relationship. While, as in the case of initial return decisions, 'mathematics and natural science,

Table 4.28 Initial return—bachelor degree discipline, (%) students

	n	Bachelor degree	
		Natural science	Social science
Return	52	47.4	56.8
Undecided	37	35.1	38.6
Stay	12	17.5	4.6
Total	101		

Chi2(2) = 4.047

Table 4.29 Initial return—bachelor degree discipline, (%) professionals

	n	Bachelor degree	
		Natural science	Social science
Return	40	47.1	48.5
Undecided	26	33.3	27.3
Stay	18	19.6	24.2
Total	84		

Chi2(2) = 0.447

Table 4.30 Initial return—bachelor degree discipline (%) (students + professionals)

	Bachelor degree	
	Natural science	Social science
Return	47.2	53.3
Undecided	34.3	33.8
Stay	18.5	13.0
n	108	77

Chi2(2) = 1.179

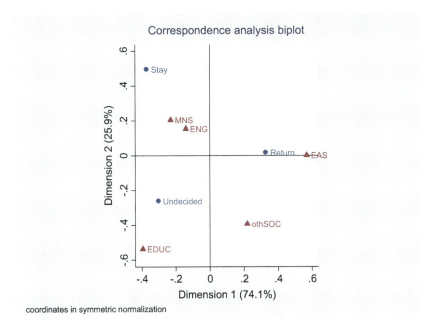

Fig. 4.6 Initial return intention—bachelor degree (all sample)

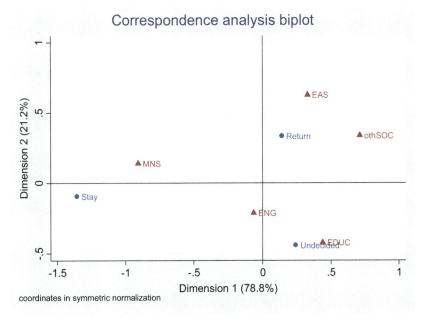

Fig. 4.7 Initial return intention—bachelor degree (students)

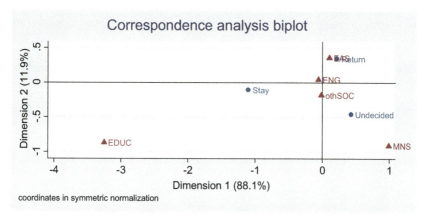

Fig. 4.8 Initial return intention—bachelor degree (professionals)

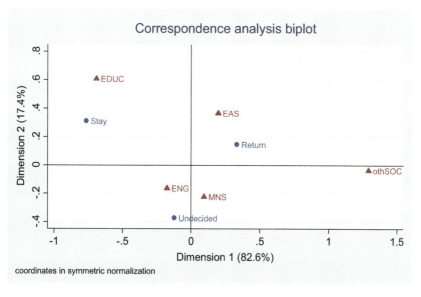

Fig. 4.9 Initial return intention—bachelor degree (male)

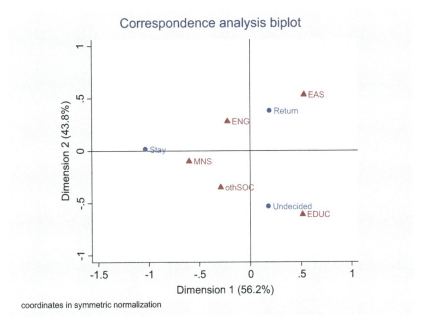

Fig. 4.10 Initial return intention—bachelor degree (female)

Table 4.31 Current return—field of study, (%) students

	n	Field of study	
		Natural science	Social science
Return immediately after studies	30	19.1	34.0
Definitely return but not soon after studies	24	30.2	9.4
Return probable	37	25.4	39.6
Return unlikely	22	23.8	13.2
Definitely not return	3	1.6	3.8
Total	116		

Chi2(4) = 12.515**; ** refers to significance at the 5% level

Table 4.32 Current return—field of study/work, (%) professionals

	n	Field of study	
		Natural science	Social science
Definitely return, plans	3	0.0	7.0
Definitely return, no plans	2	0.0	4.7
Return probable	33	48.8	30.2
Return unlikely	34	34.2	46.5
Definitely not return	12	17.1	11.6
Total	84		

Chi2(4) = 7.834*; * refers to significance at the 10% level

Table 4.33 Current return—field of study/work (%) (students + professionals)

	Field of study	
	Natural science	Social science
Definitely return, plans	11.5	21.9
Definitely return, no plans	18.3	7.3
Return probable	34.6	35.4
Return unlikely	27.9	28.1
Definitely not return	7.7	7.3
n	104	96

Chi2(4) = 7.881*; * refers to significance at the 10% level

medical health majors' are likely to stay and 'educational science', 'engineering-technical-city plan architect', and 'other social science' are more less likely to stay.

Tables 4.34, 4.35, and 4.36 show the relationship between the initial return decision and the field of study/work. Accordingly, compared to social science students, natural science students are more likely to stay, significantly at 10 percent (Table 4.34). Figures 4.16, 4.17, 4.18, 4.19, and 4.20 show that those who study/work in the field of 'economics and administrative science' are more likely to return, and those in 'mathematics and natural science, medical health' field are more likely to stay.

Tables 4.37, 4.38, 4.39, 4.40, 4.41, 4.42, and 4.43 present the results of the linkage between current (and initial) return decisions and having previous experience abroad. Although there is not significant relationship

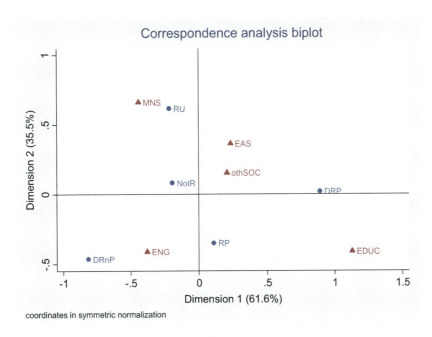

Fig. 4.11 Current return intention—field of study (all sample)

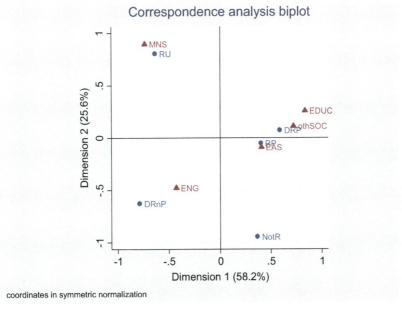

Fig. 4.12 Current return intention—field of study (students)

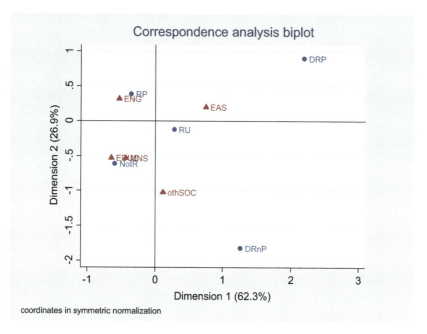

Fig. 4.13 Current return intention—field of study (professionals)

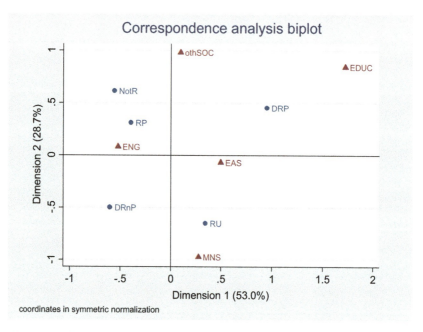

Fig. 4.14 Current return intention—field of study (male)

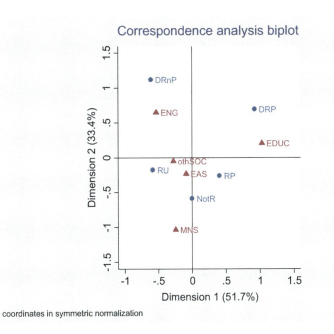

Fig. 4.15 Current return intention—field of study (female)

Table 4.34 Initial return—field of study, (%) students

	n	Field of study	
		Natural science	Social science
Return	61	47.6	58.5
Undecided	39	31.8	35.9
Stay	16	20.6	5.7
Total	116		

Chi2(2) = 5.471*; * refers to significance at the 10% level

between the initial return decision and having previous work/study/travel experience abroad, there is a highly significant linkage between having experience and the current return intention. Those who spend time abroad previously are more likely not to return.

Table 4.35 Initial return—field of study, (%) professionals

	n	Field of study	
		Natural science	Social science
Return	40	48.8	46.5
Undecided	26	34.2	27.9
Stay	18	17.1	25.6
Total	84		

Chi2(2) = 0.996

Table 4.36 Initial return—field of study (%) (students + professionals)

	Field of study	
	Natural science	Social science
Return	48.1	53.1
Undecided	32.7	32.3
Stay	19.2	14.6
n	104	96

Chi2(2) = 0.888

Tables 4.43, 4.44, 4.45, and 4.46 show that there is not substantial difference between professionals in natural science and professionals in social sciences in terms of their emphasis on certain push/pull factors, except for two cases: While 'greater job availability in specialization' is significantly emphasized more by professionals in natural science, 'economic instability, uncertainty' is significantly emphasized more by professionals in social sciences. In the case of students on the other hand, the differences between natural and social science in terms of their emphasis on certain push/pull factors are more substantial. Students of natural science put significantly more emphasis on push factors of 'low income in my occupation', 'little opportunity for advancement in occupation', and 'limited job opportunity in specialty' and on pull factors of 'higher salary or wage', 'greater advancement opportunity in profession', and 'greater job availability in specialization'.

One can have a similar observation through Tables 4.47 and 4.48. Those who have previous work/study/travel experience abroad significantly put more emphasis on aforementioned push and pull factors compared to those without work/study/travel experience abroad.

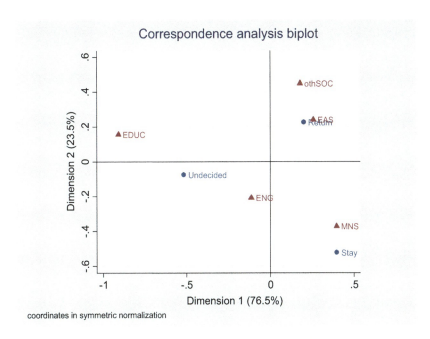

Fig. 4.16 Initial return intention—field of study (all sample)

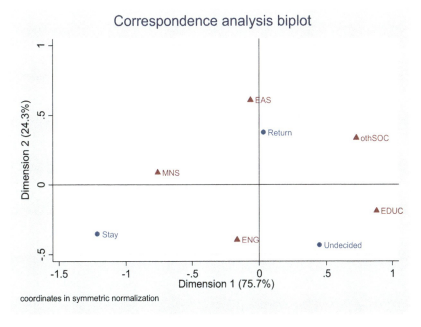

Fig. 4.17 Initial return intention—field of study (students)

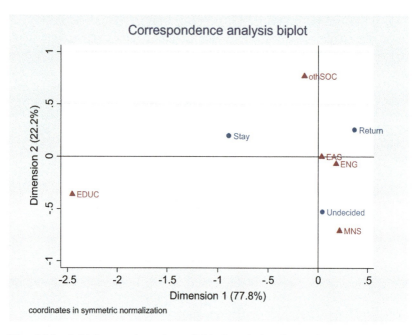

Fig. 4.18 Initial return intention—field of study (professionals)

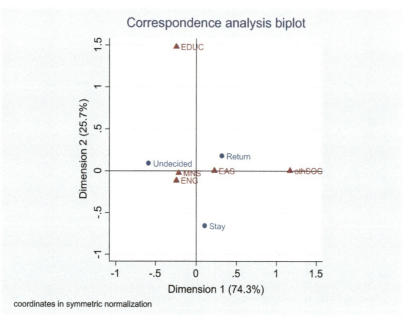

Fig. 4.19 Initial return intention—field of study (male)

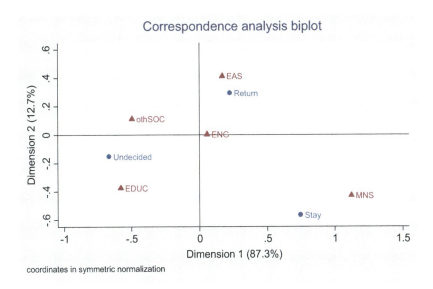

Fig. 4.20 Initial return intention—field of study (female)

Table 4.37 Current return—having previous experience (study, work, travel) abroad, (%) students

	n	Previous experience abroad	
		Yes	No
Return immediately after studies	30	19.8	40.0
Definitely return but not soon after studies	24	18.5	25.7
Return probable	37	39.5	14.3
Return unlikely	22	21.0	14.3
Definitely not return	3	1.2	5.7
Total	116		

Chi2(4) = 11.835**; ** refers to significance at the 5% level

Figures 4.21, 4.22, 4.23, 4.24, and 4.25 show the relationship between the initial return decision, the current return decision, and the stay duration in a more concise way. That is, the significant positive relationship between the initial and the current return decisions and the significant positive linkage between the current return decision and time spent abroad

Table 4.38 Current return—having previous experience (study, work, travel) abroad, (%) professionals

	n	Previous experience abroad	
		Yes	No
Definitely return, plans	3	4.9	0.0
Definitely return, no plans	2	1.6	4.4
Return probable	33	37.7	43.5
Return unlikely	34	42.6	34.8
Definitely not return	12	13.1	17.4
Total	84		

Chi2(4) = 2.255

Table 4.39 Current return—having previous experience (study, work, travel) abroad (%) (students + professionals)

	Previous experience abroad	
	Yes	No
Definitely return, plans	13.4	24.1
Definitely return, no plans	11.3	17.2
Return probable	38.7	25.9
Return unlikely	30.3	22.4
Definitely not return	6.3	10.3
n	142	58

Chi2(4) = 7.760*; * refers to significance at the 10% level

Table 4.40 Initial return—having previous experience (study, work, travel) abroad, (%) students

	n	Previous experience abroad	
		Yes	No
Return	61	49.4	60.0
Undecided	39	33.3	34.3
Stay	16	17.3	5.7
Total	116		

Chi2(2) = 2.902

Table 4.41 Initial return—having previous experience (study, work, travel) abroad, (%) professionals

	n	Previous experience abroad	
		Yes	No
Return	40	44.3	56.5
Undecided	26	31.2	30.4
Stay	18	24.6	13.0
Total	84		

Chi2(2) = 1.569

Table 4.42 Initial return—having previous experience (study, work, travel) abroad (%) (students + professionals)

	Previous experience abroad	
	Yes	No
Return	47.2	58.6
Undecided	32.4	32.8
Stay	20.4	8.6
n	142	58

Chi2(2) = 4.442

are demonstrated in Figs. 4.21, 4.22, 4.23, 4.24, and 4.25. In other words, those with initial stay decision are likely to express their current intention to stay, and this decision becomes stronger as time passes.

What is the most remarkable fact in Table 4.49 is the difference between emphasis of female and male students, particularly in the case of push factors. Table 4.49 presents two major findings: First, while the difference in the importance of push/pull factors between male and female professionals is not significant (except for two cases), there is a highly significant differentiation in the case of students. Second, particularly for students, the differentiation is much more clear among push factors than pull factors.

The findings in Tables 4.49 and 4.50 deserve a closer look as there is a clear overlap between the factors that women emphasize statistically significantly more than men and the gender gap in Turkey's labor market, particularly regarding 'little opportunities for advancement in occupation', 'limited job opportunities in specialty', 'no opportunity for advanced

Table 4.43 Push and pull factors viewed as important[a] (%) bachelor degree discipline

	Students			Professionals		
	Natural science	Social science	Chi2(1)	Natural science	Social science	Chi2(1)
Push						
Low income in my occupation	73.7	48.7	6.231**	64.7	58.1	0.362
Little opportunities for advancement in occupation	84.2	64.1	5.139**	68.6	58.1	0.941
Limited job opportunities in specialty	80.7	53.9	7.922***	68.6	58.1	0.941
No opportunity for advanced training	59.7	48.7	1.119	49.0	38.7	0.828
Away from research centers and advances	77.2	66.7	1.299	56.9	51.6	0.215
Lack of financial resources for business	38.6	30.8	0.620	39.2	22.6	2.416
Less than satisfying social/cultural life	33.3	46.2	1.607	33.3	41.9	0.615
Bureaucracy, inefficiencies	84.2	71.8	2.164	80.4	77.4	0.104
Political pressures, discord	79.0	76.9	0.056	86.3	87.1	0.011
Lack of social security	57.9	56.4	0.021	54.9	67.7	1.322
Economic instability, uncertainty	68.4	76.9	0.828	80.4	90.3	1.425
Pull						
Higher salary or wage	75.4	53.9	4.858**	74.5	64.5	0.930
Greater advancement opportunities in profession	94.7	71.8	9.784***	84.3	74.2	1.257
Better work environment	73.7	66.7	0.552	70.6	74.2	0.124
Greater job availability in specialization	84.2	66.7	4.035**	82.3	58.1	5.794**
Greater opportunities to develop specialty	89.5	79.5	1.854	82.4	67.7	2.312
More organized, ordered environment	70.2	74.4	0.201	80.4	80.6	0.000

(*continued*)

Table 4.43 (continued)

	Students			Professionals		
	Natural science	Social science	Chi2(1)	Natural science	Social science	Chi2(1)
More satisfying social/cultural life	43.9	46.2	0.050	39.2	58.1	2.755*
Proximity to research and innovation centers	77.2	74.4	0.102	62.8	58.1	0.178
Spouse's preference or job	33.3	33.3	0.000	45.1	54.8	0.732
Better educational opportunities for children	38.6	43.6	0.239	68.6	61.3	0.462
Need to finish/continue with current project	31.6	38.5	0.486	25.5	35.5	0.930
n	57	39		51	31	

*, ** and *** refer to significance at the 10%, 5% and 1% levels, respectively
[a] Marked as 'Very Important' or 'Important' by respondents

training', 'less satisfying social/cultural life', 'political pressures, discord', and 'economic instability, uncertainty'.

Regarding pull factors, it is striking, but perhaps not unexpected, that while 'better educational opportunities for children' are an important pull factor for only 25.8 percent of male students, it is considered important by 60 percent of female students.

Finally, a regression analysis is utilized to test if being female affected return intentions. We investigate what factors increase the likelihood of 'current return intention' by employing an ordered logit regression. We specified two models. In Model 1, explanatory variables are 'being female', 'age', 'square of age', 'initial return intention (undecided or stay)', 'stay duration', 'family support', and 'being professional'. Model 2 includes all push and pull factors in addition to all variables used in Model 1.

The results presented in Table 4.51 suggest a weak association between being female and non-return intention. According to Model 1, initially expressing intention to stay, being initially undecided, and having support from the family increase the likelihood of non-return intentions. Similarly, being professional, being female, staying abroad for longer, and being in a higher age group also increase non-return intentions. Model 2, which included all pull and push factors, indicated that being female and staying

Table 4.44 Push and pull factors viewed as important[a] (%) bachelor degree discipline (students + professionals)

	Natural science	Social science	Chi2(1)
Push			
Low income in my occupation	69.4	52.9	5.009**
Little opportunities for advancement in occupation	76.9	61.4	4.886**
Limited job opportunities in specialty	75.0	55.7	7.191***
No opportunity for advanced training	54.6	44.3	1.818
Away from research centers and advances	67.6	60.0	1.071
Lack of financial resources for business	38.9	27.1	2.601
Less than satisfying social/cultural life	33.3	44.3	2.171
Bureaucracy, inefficiencies	82.4	74.3	1.701
Political pressures, discord	82.4	81.4	0.028
Lack of social security	56.5	61.4	0.430
Economic instability, uncertainty	74.1	82.9	1.881
Pull			
Higher salary or wage	75.0	58.6	5.316**
Greater advancement opportunities in profession	89.8	72.9	8.716***
Better work environment	72.2	70.0	0.103
Greater job availability in specialization	83.3	62.9	9.569***
Greater opportunities to develop specialty	86.1	74.3	3.933**
More organized, ordered environment	75.0	77.1	0.106
More satisfying social/cultural life	41.7	51.4	1.632
Proximity to research and innovation centers	70.4	67.1	0.207
Spouse's preference or job	38.9	42.9	0.278
Better educational opportunities for children	52.8	51.4	0.031
Need to finish/continue with current project	28.7	37.1	1.389
n	108	70	

** and *** refer to significance at the 5% and 1% levels, respectively
[a]Marked as 'Very Important' or 'Important' by respondents

abroad for longer time do not significantly increase the probability of non-return intentions. All other significant variables in Model 1 were significant in Model 2 as well. Model 2 shows that those who emphasize the push factors of low income in one's occupation and bureaucracy and inefficiencies and the pull factors of greater job availability in specialization and better educational opportunity for children are more likely to express non-return intentions.

Table 4.45 Push and pull factors viewed as important[a] (%) field of study

	Students			Professionals		
	Natural science	Social science	Chi2(1)	Natural science	Social science	Chi2(1)
Push						
Low income in my occupation	69.8	50.0	4.519**	65.9	58.5	0.467
Little opportunities for advancement in occupation	79.4	66.7	2.276	65.9	63.4	0.053
Limited job opportunities in specialty	74.6	56.3	4.130**	68.3	60.1	0.480
No opportunity for advanced training	55.6	50.0	0.338	43.9	46.3	0.049
Away from research centers and advances	74.6	64.6	1.309	58.5	51.2	0.443
Lack of financial resources for business	36.5	33.3	0.121	46.3	19.5	6.682**
Less than satisfying social/cultural life	31.8	45.8	2.299	31.7	41.5	0.841
Bureaucracy, inefficiencies	79.4	68.8	1.628	75.6	82.9	0.668
Political pressures, discord	74.6	72.9	0.040	85.4	87.8	0.105
Lack of social security	55.6	58.3	0.086	56.1	63.4	0.456
Economic instability, uncertainty	63.5	77.1	2.368	75.6	92.7	4.479**
Pull						
Higher salary or wage	73.0	58.3	2.643*	73.2	68.3	0.236
Greater advancement opportunities in profession	92.0	72.9	7.35***	85.4	75.6	1.242
Better work environment	71.4	68.8	0.094	70.8	73.2	0.060
Greater job availability in specialization	80.1	66.7	2.948*	82.9	63.4	3.976**
Greater opportunities to develop specialty	84.1	79.2	0.454	85.4	68.3	3.357*
More organized, ordered environment	66.7	77.1	1.441	78.1	82.9	0.311
More satisfying social/cultural life	42.9	45.8	0.098	43.9	48.8	0.196

(*continued*)

Table 4.45 (continued)

	Students			Professionals		
	Natural science	Social science	Chi2(1)	Natural science	Social science	Chi2(1)
Proximity to research and innovation centers	71.4	75.0	0.176	63.4	58.5	0.205
Spouse's preference or job	31.8	31.3	0.003	43.9	53.7	0.781
Better educational opportunities for children	36.5	43.8	0.597	68.3	63.4	0.217
Need to finish/continue with current project	28.6	39.6	1.487	24.4	34.2	0.943
n	63	48		41	41	

*, ** and *** refer to significance at the 10%, 5% and 1% levels, respectively
[a]Marked as 'Very Important' or 'Important' by respondents

Table 4.46 Push and pull factors viewed as important[a] (%) field of study (students + professionals)

	Natural science	Social science	Chi2(1)
Push			
Low income in my occupation	68.3	53.9	4.170**
Little opportunities for advancement in occupation	74.0	65.2	1.795
Limited job opportunities in specialty	72.1	58.4	3.993**
No opportunity for advanced training	51.0	48.3	0.134
Away from research centers and advances	68.3	58.4	2.010
Lack of financial resources for business	40.4	27.0	3.837**
Less than satisfying social/cultural life	31.7	43.8	2.997*
Bureaucracy, inefficiencies	77.9	75.3	0.182
Political pressures, discord	78.9	79.8	0.252
Lack of social security	55.8	60.7	0.474
Economic instability, uncertainty	68.3	84.3	6.665***
Pull			
Higher salary or wage	73.1	62.9	2.288
Greater advancement opportunities in profession	89.4	74.2	7.701***
Better work environment	71.2	70.8	0.003
Greater job availability in specialization	81.7	65.2	6.854***
Greater opportunities to develop specialty	84.6	74.2	3.253*

(*continued*)

Table 4.46 (continued)

	Natural science	Social science	Chi2(1)
More organized, ordered environment	71.2	79.8	1.908
More satisfying social/cultural life	43.3	47.2	0.298
Proximity to research and innovation centers	68.3	67.4	0.016
Spouse's preference or job	36.5	41.6	0.512
Better educational opportunities for children	49.0	52.8	0.273
Need to finish/continue with current project	26.9	37.1	2.288
n	*104*	*89*	

*, ** and *** refer to significance at the 10%, 5% and 1% levels, respectively
ᵃMarked as 'Very Important' or 'Important' by respondents

Table 4.47 Push and pull factors viewed as important[a] (%) having previous experience (study, work, travel) abroad

	Students			Professionals		
	No	Yes	Chi2(1)	No	Yes	Chi2(1)
Push						
Low income in my occupation	43.8	68.4	5.810**	52.2	66.1	1.365
Little opportunities for advancement in occupation	62.5	78.5	3.014*	47.8	71.2	3.951**
Limited job opportunities in specialty	59.4	69.6	1.076	52.2	69.5	2.171
No opportunity for advanced training	59.4	50.6	0.699	34.8	49.2	1.380
Away from research centers and advances	68.8	70.9	0.050	39.1	61.0	3.201*
Lack of financial resources for business	37.5	34.2	0.110	34.8	32.2	0.050
Less than satisfying social/cultural life	34.4	39.2	0.229	39.1	35.6	0.089
Bureaucracy, inefficiencies	68.8	77.2	0.865	82.6	78.0	0.217
Political pressures, discord	75.0	73.4	0.030	82.6	88.1	0.435
Lack of social security	56.3	57.0	0.005	56.5	61.0	0.139
Economic instability, uncertainty	62.5	72.2	0.999	87.0	83.1	0.189
Pull						
Higher salary or wage	53.1	72.2	3.710*	69.6	71.2	0.021
Greater advancement opportunities in profession	71.9	88.6	4.693**	82.6	79.7	0.092
Better work environment	56.3	76.0	4.231**	65.2	74.6	0.718
Greater job availability in specialization	62.5	79.8	3.592*	69.6	74.6	0.212
Greater opportunities to develop specialty	75.0	84.8	1.484	69.6	79.7	0.948

(*continued*)

Table 4.47 (continued)

	Students			Professionals		
	No	Yes	Chi2(1)	No	Yes	Chi2(1)
More organized, ordered environment	62.5	74.7	1.648	73.9	83.1	0.880
More satisfying social/cultural life	46.9	43.0	0.136	34.8	50.9	1.718
Proximity to research and innovation centers	59.4	78.5	4.215**	47.8	66.1	2.323
Spouse's preference or job	25.0	34.2	0.889	56.5	45.8	0.767
Better educational opportunities for children	43.8	38.0	0.318	60.9	67.8	0.353
Need to finish/continue with current project	34.4	32.9	0.022	17.4	33.9	2.178
n		79	32			

* and ** refer to significance at the 10% and 5% levels, respectively

[a]Marked as 'Very Important' or 'Important' by respondents

Table 4.48 Push and pull factors viewed as important[a] (%) having previous experience (study, work, travel) abroad (students + professionals)

	No	Yes	Chi2(1)
Push			
Low income in my occupation	47.3	67.4	6.733***
Little opportunities for advancement in occupation	56.4	75.4	6.753***
Limited job opportunities in specialty	56.4	69.6	3.046*
No opportunity for advanced training	49.1	50.0	0.013
Away from research centers and advances	56.4	66.7	1.806
Lack of financial resources for business	36.7	33.3	0.161
Less than satisfying social/cultural life	36.4	37.7	0.029
Bureaucracy, inefficiencies	74.6	77.5	0.197
Political pressures, discord	78.2	79.7	0.056
Lack of social security	56.4	58.7	0.088
Economic instability, uncertainty	72.7	76.8	0.356
Pull			
Higher salary or wage	60.0	71.7	2.507
Greater advancement opportunities in profession	76.4	84.8	1.921
Better work environment	60.0	75.4	4.506**
Greater job availability in specialization	65.5	77.5	2.991*
Greater opportunities to develop specialty	72.7	82.6	2.382
More organized, ordered environment	67.3	78.3	2.541
More satisfying social/cultural life	41.8	46.4	0.330
Proximity to research and innovation centers	54.6	73.2	6.269**
Spouse's preference or job	38.2	39.1	0.015
Better educational opportunities for children	49.1	49.3	0.001
Need to finish/continue with current project	27.3	33.3	0.668
n	55	138	

*, ** and *** refer to significance at the 10%, 5% and 1% levels, respectively

[a]Marked as 'Very Important' or 'Important' by respondents

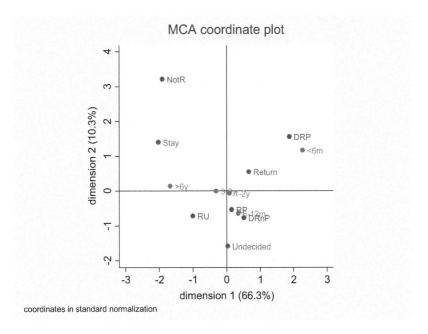

Fig. 4.21 Initial-current return intentions—stay duration (all sample)

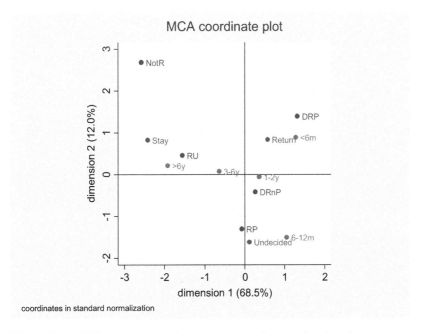

Fig. 4.22 Initial-current return intentions—stay duration (students)

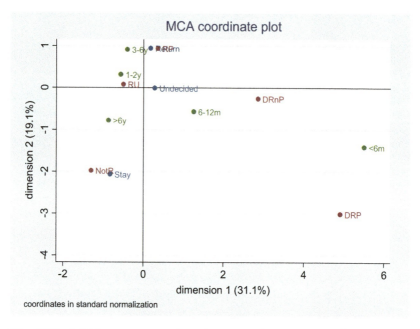

Fig. 4.23 Initial-current return intentions—stay duration (professionals)

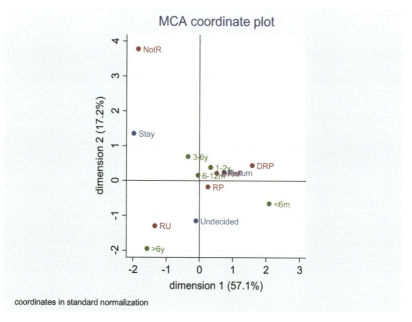

Fig. 4.24 Initial-current return intentions—stay duration (male)

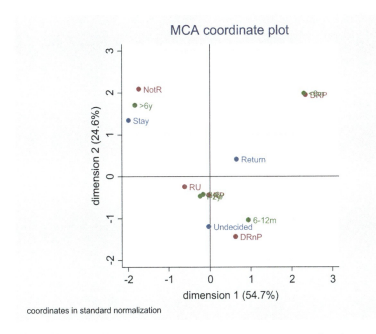

Fig. 4.25 Initial-current return intentions—stay duration (female)

Table 4.49 Push and pull factors viewed as important[a] (%)

	Students			Professionals		
	Male	Female	Chi2(1)	Male	Female	Chi2(1)
Push						
Low income in my occupation	57.6	66.7	0.932	61.1	64.3	0.079
Little opportunities for advancement in occupation	65.2	86.7	6.418**	55.6	82.1	5.702**
Limited job opportunities in specialty	60.6	75.6	2.691	59.3	75.0	1.999
No opportunity for advanced training	43.9	66.7	5.556**	37.0	60.7	4.177**
Away from research centers and advances	63.6	80.0	3.430*	53.7	57.1	0.088
Lack of financial resources for business	30.3	42.2	1.668	35.2	28.6	0.365

(*continued*)

Table 4.49 (continued)

	Students			Professionals		
	Male	Female	Chi2(1)	Male	Female	Chi2(1)
Less than satisfying social/cultural life	30.3	48.9	3.930**	35.2	39.3	0.135
Bureaucracy, inefficiencies	71.2	80.0	1.096	79.6	78.6	0.013
Political pressures, discord	66.7	84.4	4.382**	85.2	89.3	0.267
Lack of social security	48.5	68.9	4.539**	61.1	57.1	0.121
Economic instability, uncertainty	58.1	84.4	8.094***	81.5	89.3	0.842
Pull						
Higher salary or wage	68.2	64.4	0.168	66.7	78.6	1.262
Greater advancement opportunities in profession	80.3	88.9	1.452	77.8	85.7	0.740
Better work environment	65.5	77.8	2.042	66.7	82.1	2.188
Greater job availability in specialization	72.7	77.8	0.362	70.4	78.6	0.632
Greater opportunities to develop specialty	77.2	88.9	2.444	79.6	71.4	0.697
More organized, ordered environment	65.2	80.0	2.875*	77.8	85.7	0.740
More satisfying social/cultural life	37.9	53.3	2.592	50.0	39.3	0.851
Proximity to research and innovation centers	68.2	80.0	1.895	59.3	64.3	0.196
Spouse's preference or job	28.8	35.6	0.568	48.2	50.0	0.025
Better educational opportunities for children	25.8	60.0	13.112***	64.8	67.9	0.076
Need to finish/continue with current project	25.8	44.4	4.205**	31.5	25.0	0.374
n	66	45		54	28	

*, ** and *** refer to significance at the 10%, 5% and 1% levels, respectively

[a]Marked as 'Very Important' or 'Important' by respondents

Table 4.50   Push and pull factors rated as 'Important' or 'Very Important' (%) (students + professionals)

	Male	Female	Chi-square
Push			
Low income in my occupation	59.2	65.8	0.833
Little opportunities for advancement in occupation	60.8	84.9	12.539***
Limited job opportunities in specialty	60.0	75.3	4.748**
No opportunity for advanced training	40.8	64.4	10.070***
Away from research centers and advances	59.2	71.2	2.859*
Lack of financial resources for business	32.5	37.0	0.406
Less satisfying social/cultural life	32.5	45.2	3.133*
Bureaucracy, inefficiencies	75.0	79.5	0.503
Political pressures, discord	75.0	86.3	3.528*
Lack of social security	54.2	64.4	1.945
Economic instability, uncertainty	69.2	86.3	7.234***
Pull			
Higher salary or wage	67.5	69.9	0.117
Greater advancement opportunity in profession	79.2	87.7	2.262
Better work environment	65.8	79.5	4.087**
Greater job availability in specialization	71.2	78.1	0.973
Greater opportunities to develop specialty	79.3	82.2	0.419
More organized, ordered environment	70.8	82.2	3.134*
More satisfying social/cultural life	43.3	48.0	0.390
Proximity to research and innovation centers	64.2	74.0	2.002
Spouse's preference or job	37.5	41.1	0.247
Better educational opportunities for children	43.3	63.0	7.034***
Need to finish/continue with current project	28.3	37.0	1.572
n	120	73	

Note: *, **, and *** refer to significance at the 10 percent, 5 percent, and 1 percent levels, respectively

Overall, these findings reveal several differences between men and women in terms of their return intentions (Elveren and Toksöz 2017), mostly supporting early findings of Güngör (2014), a research based on a survey conducted in 2002. First, current non-return intentions are significantly higher than initial non-return intentions for women. Women are more likely to appreciate living abroad due to the greater opportunities there or the lack of opportunities in their home country (Elveren and Toksöz 2017).

The second finding is that females who reported their intention to return are more likely to go abroad again compared to the same group of males. Third, there is remarkable overlap with findings of Güngör (2014)

Table 4.51 Non-return intention, regression results

No	Model 1	Model 2
Female	0.497* (0.280)	0.369 (0.332)
Age	0.399** (0.180)	0.444** (0.202)
Age2	−0.005** (0.003)	−0.006** (0.003)
Initial return decision (undecided)	1.021*** (0.303)	1.351*** (0.347)
Initial return decision (stay)	1.990*** (0.439)	2.465*** (0.484)
Stay duration	0.087** (0.042)	0.059 (0.049)
Family support	0.514*** (0.148)	0.509*** (0.174)
Professional	0.765* (0.411)	1.013** (0.447)
Push factors		
Low income in my occupation		0.960** (0.440)
Little opportunities for advancement in occupation		−0.259 (0.520)
Limited job opportunities in specialty		−0.457 (0.438)
No opportunity for advanced training		0.028 (0.438)
Away from research centers and advances		0.124 (0.452)
Lack of financial resources for business		−0.251 (0.341)
Less satisfying social/cultural life		0.341 (0.374)
Bureaucracy, inefficiencies		0.771* (0.459)
Political pressures, discord		0.332 (0.544)
Lack of social security		−0.572 (0.388)
Economic instability, uncertainty		0.165 (0.482)
Pull factors		
Higher salary or wage		−0.242 (0.437)
Greater advancement opportunity in profession		0.763 (0.592)
Better work environment		−0.241 (0.442)
Greater job availability in specialization		1.080** (0.489)
Greater opportunities to develop specialty		−0.128 (0.540)
More organized, ordered environment		−0.299 (0.446)
More satisfying social/cultural life		0.233 (0.380)
Proximity to research and innovation centers		0.025 (0.413)
Spouse's preference or job		−0.133 (0.346)
Better educational opportunities for children		0.692* (0.374)
Need to finish/continue with current project		−0.486 (0.358)
Observations	199	192

Note: *, **, and *** refer to significance at the 10 percent, 5 percent, and 1 percent levels, respectively in that women emphasize the gender gap in labor market in Turkey as main push factors in their migration decision, such as 'little opportunities for advancement in occupation', 'limited job opportunities in specialty' and 'no opportunity for advanced training'. Also, as noted in Güngör (2014), women are more likely to migrate and not return because they are more affected by economic downturns and instability.

There are, however, two striking differences with Güngör (2014). While 'political pressures, discord' and 'better educational opportunities for children' are emphasized more by men, we found the opposite results. There is no plausible explanation for the different findings on better future for children as a pull factor except for selection bias problem. On the other hand, one can argue that the reason why women, not men, emphasized 'political pressures, discord' more in this follow-up study is the changing characteristics of 'political pressure' that affect women's lives more seriously. That is what the next chapter will examine: this new characteristic of the political environment, namely, Islamic authoritarianism.

Notes

1. Questionnaire is based on Güngör (2003).
2. This might be due partly to the snowball method.

References

Elveren, A. Y., & Toksöz, G. (2017). Why Don't Highly Skilled Women Want to Return? Turkey's Brain Drain from a Gender Perspective. MPRA No. 80290, 2017. https://mpra.ub.uni-muenchen.de/80290/

Gökbayrak, Ş. (2009). Skilled Labour Migration and Positive Externality: The Case of Turkish Engineers Working Abroad. *International Migration*, 50(s1), 132–150.

Güngör, N. D. (2003). *Brain Drain from Turkey: An Empirical Investigation of the Determinants of Skilled Migration and Student Non-return.* Unpublished PhD dissertations, Department of Economics, Middle East Technical University, Ankara, Turkey.

Güngör, N. D. (2014). *The Gender Dimension of Skilled Migration and Return Intentions.* Invited presentation, Harnessing knowledge on the migration of highly skilled women: An expert group meeting, International Organization for Migration and OECD Development Center, Geneva, Switzerland, April 3–4, 2014.

Güngör, N. D., & Tansel, A. (2008a). Brain Drain from Turkey: An Investigation of Students' Return Intention. *Applied Economics*, 40, 3069–3087.
Güngör, N. D., & Tansel, A. (2008b). Brain Drain From Turkey: The Case of Professionals Abroad. *International Journal of Manpower*, 29, 323–347.
Güngör, N. D., & Tansel, A. (2014). Brain Drain From Turkey: Return Intentions of Skilled Migrants. *International Migration*, 52(5), 208–226.

CHAPTER 5

The Transformation of Regime and Gender Inequality in Turkey

Abstract The common perception in the West was that, with the AKP, Turkey had finally overcome the enduring Islamist versus secularist societal split, and it was believed that 'New Turkey' of the Muslim democratic AKP could be a role model for the Arab World. However, history has proved correct the alternative narrative by Turkey's secular elite that the AKP or Erdoğan's hidden agenda is to replace Kemalist ideology with an authoritarian Islamic one. This chapter provides a brief account of this radical transformation, focusing on the welfare regime and education system.

Keywords Justice and Development Party • Welfare state • Authoritarianism • Islamic charity • Gender • Religious education

Women's empowerment was a key element of the modernization project of the new Republican Turkey based on the vision of its founder, Mustafa Kemal Atatürk. Creating women's equality in the public sphere was a core national policy. Although the project was unable to reach out to rural women, overall it is generally considered a great success as the nation radically transformed itself from the ruins of the Ottoman Empire into a Westernized state. The regime has radically changed during the AKP era toward an authoritarian Islamic regime.

© The Author(s) 2018
A. Y. Elveren, *Brain Drain and Gender Inequality in Turkey*,
https://doi.org/10.1007/978-3-319-90860-1_5

Karl Polanyi argues that capitalist development involves two simultaneous and contradictory tendencies: the extension of the scope of the market, which has detached itself from society and the countermovement of different classes and organization in society to protect themselves from this change (Polanyi 1944). This 'double movement' helps to explain 'the great transformation' in Turkey during the AKP era (Eres 2011). The AKP came to power after the socio-economic situation reached its lowest point during the 2001 economic crisis. The despair of society made the AKP a savior in the eyes of the poor. However, what makes this situation unique is that the AKP continued the very same paradigm that had created the despair itself while using Islamic forms of charity very effectively to reach out to the most vulnerable segment of society. This development, a daring merger between neoliberalism and Islamic authoritarianism, so-called Islamic neoliberalism, has been widely discussed (inter alia Acar and Altunok 2013; Buğra 2012, 2013; Dedeoğlu 2013). This new approach, which assigns welfare provision to non-state actors, was perfectly in line with the insight of international institutions in fighting against poverty. Simultaneously, it was a powerful tool for the AKP to mobilize the poor—particularly poor women—in order to expand its vote base and extend the Islamic and conservative values in both the public and private spheres via reforms in welfare provision and education.

Women became key actors in this transformation, both as active party supporters and as the victims of the transformation. It is true that this unique combination created by the AKP benefitted poor (and) conservative women by providing social aid in both cash and in kind while simultaneously lessening their confinement to the private sphere by giving them more opportunities to enjoy the public sphere. It is also true that some reforms to Turkey's civil and penal codes benefitted all women to some extent, particularly upper-class women who are already in the labor market. However, as discussed in this section, women are also hurt more by this transformation because the new regime sees women socially subordinate.

5.1 Pre-AKP Period

This section discusses three main problems that pave the way for AKP to come to power, namely, economic fault line, the Kurdish problem, and rising political Islam[1] (Özden et al. 2017).

5.1.1 An Economic Fault Line

The 1980 military coup, the breakpoint in the formation of contemporary Turkey, facilitated the establishment of the neoliberal paradigm by repressing the voice of civil society and shutting down the largest labor union. The military government's civilian successor followed the same neoliberal model. However, this transformation in Turkey's welfare regime toward a market-oriented economic paradigm intensified most after AKP came to power in 2002. What distinguishes this era (which has continued since then) is not just the government's adherence to the boldest neoliberal agenda in the history of the Turkish Republic but rather how they created a unique system through a predator state and Islamization, which I name *the Pious Predator State* (to be discussed later in the section).

The Turkish economy in the post-1980 era was characterized by increased volatility in economic growth, persistently high inflation, expanding fiscal deficits, marginalization of the labor force, and increasing poverty and income inequality (Yeldan 2006).

The main characteristic of the 1981–1988 subperiod was export promotion in a regulated foreign exchange system and controlled capital inflows (Boratav et al. 2002; Yeldan 2006). During this period, Turkey's economy was integrated into global markets through commodity trade liberalization. The exchange rate and direct export subsidies played key roles in promoting exports and sustaining macroeconomic stability. Until populist pressures from approaching elections encouraged significant increases, real wages remained stagnant during this period because the environment was hostile to organized labor (Boratav et al. 2002; Yeldan 2006). Compared to their levels in 1980, real wages in manufacturing in both the public and private sectors were stagnated at negative levels until 1988, before increasing by 91 percent and 58 percent by 1992, respectively (Elveren and Galbraith 2009). This substantial raise in real wages was accompanied by increase in investments on social infrastructure.

The second subperiod after 1980 started in 1989 when capital accounts were liberalized and the Turkish lira became fully convertible in foreign exchange markets. This allowed an inflow of short-term foreign capital, or so-called hot money, which made it possible to finance increasing public expenditure and reduce the cost of imports (Boratav et al. 2002; Yeldan 2006).

This newly deregulated financial environment increased the dependence of the Turkish economy on short-term, speculative capital. Sources of economic growth thus originated from highly volatile foreign financial capital rather than domestic capital accumulation. In fact, a sudden drainage of short-term funds at the beginning of January 1994 reduced production capacity, causing a major economic crisis. Economic growth of 8.0 percent in 1993 dropped to −5.5 percent in 1994, while inflation rose to three-digit levels, ironically when a professor of economics was the prime minister. The Asian and Russian financial crises also hit the Turkish economy, reducing economic growth from an average of 7.2 percent after the 1994 crisis to 3.1 percent in 1998 and −5.0 percent in 1999. The Central Bank implemented a program in December 1999 to reduce inflation so as to address increasing instability and ongoing inflation. This exchange rate-based program, prepared under the direct supervision of the International Monetary Fund (IMF), functioned for only a short time, and the upturn in economic activity did not last long. Instead, the economy shrunk 7.6 percent in 2001, inflation rose to 86 percent, and unemployment rose from 6.5 percent in 2000 to 8.4 percent in 2001 and to as high as 11.5 percent in the first quarter of 2002. Manufacturing real wages fell 12.2 percent in the private sector and 10.3 percent in the public sector (Elveren and Galbraith 2009).

Overall, the post-1980 period was characterized by stubbornly high unemployment rates, increasing wage inequality between manufacturing sectors, provinces, and geographical regions (Elveren and Galbraith 2009; Elveren 2010), and between formal and informal sectors (Aydin et al. 2010), suppressed real wages compared to a steady increase in labor productivity (Voyvoda and Yeldan 2001; Memis 2007a; Elgin and Kuzubaş 2012) that increased profits (Eres 2005; Memis 2007b; Ongan 2011), and a decreased wage share (Onaran 2009). The empirical literature on this period identified several causes of the increasing inequality in incomes and wages: increasing internationalization (e.g., Çağatay 1986; Boratav 1990; Kızılırmak 2003; Meschi et al. 2008; Aydıner-Avşar 2011; Oyvat 2011; Elveren et al. 2012), decreasing unionization (Elveren 2013a), increasing military expenditure (Elveren 2012), and an expanding informal sector (Elveren and Özgür 2016).

5.1.2 Two More Fault Lines: The Kurdish Problem and Rising Political Islam

Having briefly discussed the economic strains of the post-1980 period, the other fault lines to consider lie in the political sphere, namely, the Kurdish problem, and the rise of political Islam.

The Kurdish problem in Turkey is as old as Republican Turkey. There has been a low-intensity conflict between the Turkish armed forces and the terrorist organization, Kurdistan Workers' Party (PKK), since the early 1980s. It is true that socio-economic inequality between subcultures may be a breeding ground for conflict. In fact, 'nonmaterial insecurity'—that is, deficiency of language, culture, identity, and alienation (İçduygu et al. 2010)—has acted as a 'persisting factor' for (separatist) terrorism in Turkey (Derin-Güre and Elveren 2014). While the traditional response of the state in Turkey to the ongoing conflict has been primarily military, policy-makers have mainly focused on improving the region's economic conditions as conventional wisdom suggests that economic deprivation in terms of income, income inequality, poverty, and so on has promoted terrorism in Turkey. However, military measures alone have created a vicious cycle, as rightly elaborated in Derin-Güre and Elveren (2014). After the conflict intensified in the 1990s, around two million Kurdish people were expelled from the region. Most became permanent members of the marginal workforce in urban sectors, with no social insurance or benefits. As Buğra and Keyder (2006) noted, a pillar of the welfare regime, namely, enduring ties with relatives in the village that provides welfare provision, was never an option for Kurdish migrants to Turkey's cities.

The second element of the political fault line, between secularists and Islamists, also dates back to the foundation of Republican Turkey. A key aspect of Republican Turkey's original modernization project was the control of religion by state institutions. The new secular political system promoted a moderate, Turkish Islam while controlling political Islam by military-backed laws.

Until it established its first separate party in 1970, political Islam was embedded in mainstream conservative parties. After the first Islamic party was shut down during the 1971 military coup, its successor, which successfully participated in coalition governments during the 1970s, was also banned during the 1980–1983 military regime along with the other political parties. It is important to note clearly that the policies of this military regime and its civilian successor ironically paved away for the later Islamization of Turkish society, through the introduction of compulsory religious courses, the establishments of numerous Koran courses and Imam-Hatip schools—vocational schools to provide training for imams and preachers in mosques—and the creation of a welcoming environment for religious sects.

The third Islamic party, the Welfare Party (Refah Partisi, RP) founded in 1983, marked a break point in the history of political Islam in Turkey since it was the first Islamic party to govern as the dominant partner of a coalition government. The party's success was based on its ability to reach out to conservative Muslims from all segments and strata of the country and the relative Islamization of the Turkish society due to the 1980–1983 military regime's policies noted above. Public discontent with corruption in fragmented mainstream secular parties and despair regarding the intensifying Kurdish conflict also played a role in the party's success (Heper 1997; Öniş 1997).

On February 28, 1997, Turkey's military-dominated National Security Council publicly criticized the Welfare Party-led government, demanding the implementation of a number of secularist policies, which became known as the 'postmodern coup'. The party was then forced to resign from the coalition government, while the military launched a campaign against actual and perceived Islamist political and economic actors and initiated a series of reforms to undo Islamization in the education system. In 1998, the Welfare Party was banned from politics by the Constitutional Court of Turkey for its alleged anti-secular activities. Its successor, the Virtue Party (Fazilet Partisi), from which the AKP was later founded by the reformer moderates (Yenilikçiler), was also shut down in 2001 for being the continuation of the RP.

It is also widely argued that Islamist parties of the post-1980s era also grew due to another fault line related to the rise of medium-scale, Islamic capital in many Anatolian cities. The establishment of organized industrial districts as part of the state's industrialization strategy during this era encouraged Anatolian small businesses by reducing their dependency on bank owning big capital groups for access to credit. Their ever-growing influence on the sociopolitical arena became disproportionate to their limited economic capacity.

5.2 The AKP Era: The Boiling Frog

The AKP was established by centrist politicians with primarily Islamist backgrounds. Although the party initially defined itself as 'conservative democratic', it gradually became 'Islamic authoritarian'. This transformation became more noticeable following the party's victory in the 2011 election and accelerated further after the Gezi Park protests in 2013. It began with a small group of environmentalist protesters opposed to the destruction of the Gezi Park in central Istanbul. The violence used by the police led to nationwide anti-government demonstrations, an outcome of growing concern of secular side of the country over changing regime. Police brutality led to around 8000 injured people, including critical injuries, loss of sight, and 11 fatalities. Most recently, the transformation of the regime was completed with the party's highly controversial referendum victory that gives ultimate political power to President Erdoğan.

After first coming to power in the 2002 elections right after the devastating 2001 economic crisis, the AKP continued the preceding coalition government's macroeconomic strategy designed under IMF guidance. This helped to maintain the support of dominant business groups. By taking advantage of a fortunate global economic climate with an abundance of global liquidity, the government was able to stabilize the Turkish lira, reduce inflation to single digits, and generate spectacular economic growth. Furthermore, the party showed full commitment to Turkey's EU accession process while establishing a foreign policy of zero problems with neighbors. The party also took a brave step regarding the enduring Kurdish problem. Finally, using the EU accession process, the party was able to reform the National Security Council, the key organization through which the military used its power to influence decision-making.

This major political stance made the party more palatable both domestically and internationally. The West's admiration for AKP claimed that Turkey had finally overcome the enduring Islamist versus secularist societal split and that the Muslim democratic AKP was building a liberal, pluralist society, and this 'New Turkey' could be a role model for the Arab World. However, history proved that Turkey's secular elite was correct as they argued that the AKP or Erdoğan has a *hidden agenda* to replace Kemalist ideology with an authoritarian Islamic one (Elveren 2018).

'The AKP era is thus a "boiling frog" situation in Turkey, in which Islamic values have gradually come to dominate every aspect of life. This gradual increase in the temperature has been evidenced by many developments: muzzling the press, eroding the rule of law and civil liberties, desecularizing Turkey's top state institutions, making loyalty to the party—and encouragement of beards and headscarves—the sole criteria for getting a government job, the rise of the Presidency of Religious Affairs, state-supported pressure to fast during Ramadan and restrictions on alcohol in small Anatolian cities and towns, adjusting working hours according to Friday prayer times, introducing women-only buses and religiously-focused curriculum reforms' (Elveren 2018, pp. 88–89).

Erdoğan was even happy to clearly point out that the water had already boiled by declaring: 'whether one accepts it or not, Turkey's administrative system has changed. Now, what should be done is to update this de facto situation within the legal framework of the constitution.'[2]

The AKP has used various strategies to extend its base. It has generated and supported 'pious' business groups with strong ties to the party by extensively utilizing public procurement. In this sense, there is a great resemblance between what James K. Galbraith elaborates in his *The Predator State* and how the relationship between private sector and the state has evolved during the AKP era (Galbraith 2008). However, to better explain the case of Turkey I modify it as *the Pious Predator State* (Elveren 2018). In the AKP era, it was not neoliberal paradigm in the conventional sense. 'It was a growing government rather than a smaller one, and it was an abandonment of the principles of fair competition, generating a reckless coalition between the state and a pious private sector. And it was not the era of *deregulation* but rather *reregulation* of the business sphere in order to strengthen the organic ties of the coalition' (Elveren 2018, pp. 87–88).

The party has also mobilized the socially and economically marginalized strata of society by extending social protection to previously excluded individuals. In fact, it is safe to argue that public procurement was the party's key strategy for extending and strengthening its vote base. It was so crucial that the party has not hesitated to revise the law governing public procurement over 200 times since 2003. In return, these business groups have financed charitable work in support of the party, a crucial tool that the government has used extensively to provide social aid to the poor. It should be noted that the AKP's use of non-state actors in collaboration with the state's involvement in welfare provision is perfectly in line with the recommendations of the World Bank. In fact, the AKP laid down milestones of the transformation toward an authoritarian Islamic regime by providing social aid based on traditional Islamic charity.

The government first addressed structural problems in Turkey's social security system.[3] The 2006 reform fixed the fragmented structure and corrected the lack of uniformity in norms and standards that had created hierarchies among social insurance beneficiaries. Under General Health Insurance (GHI), all health insurance benefits were integrated and all citizens covered. The Social Security Institution united the three previous public insurance funds and the Green Card Scheme, which previously provided basic health care for low-income citizens. GHI became the AKP's most important 'success'. Its health policies were received with enormous gratitude by the poorest segment of society, as evident from surveys showing a substantial increase in public satisfaction with health policies from 2003 to 2010 (Karadeniz 2012).

Clientelism has become the most significant characteristics of the current welfare regime. Social aid in general and conditional cash transfers in particular have been used to strengthen and expand the party's voter base. The government made itself even more attractive by offering monthly payments to poor and single women and financial support to families of those with disabilities (Buğra and Candaş 2011). The key issue, however, is that municipalities and religious NGOs encouraged and adopted Islamic traditions of charity, replacing official state channels in providing social aid and services. They now provide social aid, according to biased targeting mechanisms with clientelistic motives rather than citizenship (Elveren and Ağartan 2017).

5.2.1 The AKP's Strongest Weapons: Islamic Charity and Conditional Cash Transfers

In addition to this transformation in the health-care sector, the AKP has also improved social assistance schemes, which have become the party's most effective policy tool, both in terms of successfully reducing poverty and being able to mobilize poor people. The government has mainly relied on faith-based charitable organizations and local municipalities to provide social assistance and alleviate poverty, with strong clientelist motives.

Transferring the state's responsibility to nongovernmental organizations (particularly to faith-based groups as research found that the poor trust them more than state institutions) was one key element of the neoliberal paradigm to alleviate social exclusion and poverty while reducing state social spending. The other key element of this restructuring was to reemphasize the central role of the family (Bedford 2008; Larner 2000; Smith 2008; Atalay 2017).

Perhaps there could not be such better combination of faith-based organization and family for a political party like the AKP. The mutually beneficial relationship between the government and Islamic charity organizations has been crystal clear. On the one hand, government ambitiously supports 'pious' business groups with strong ties to the party by extensively utilizing public procurement. Donating the Islamic charity organizations is one way of showing loyalty to the government. On the other hand, these Islamic charity organizations promote 'religious familism, gendered division of labor in private and public spheres, and pronatalist biopolitics' (Atalay 2017, p. 3).

AKP has also used the social assistance program strategically in the context of the Kurdish problem. As noted earlier the conventional policy response of the state to the ongoing conflict in Southeastern Turkey was (in addition to military measures) to improve the economic conditions in the region (Derin-Güre and Elveren 2014). The Southeast Anatolia Project that has been in effect since the 1970s has not improved the economic conditions in the region[4] (Yörük and Özsoy 2013). East-West dualism has remained because the project has not benefitted the poor in the absence of substantial land reform in the region.

In this sense, AKP pursued a different strategy. Rather than increasing investment in the region at the macro level—as a matter of fact, there has been a substantial decline in public investment to the region since AKP has come to power—the government has built a clientelist network among Kurds by providing social assistance disproportionately to the Kurdish minority not just in the Southeastern region but also in urban and metropolitan areas in general (Yörük 2012). Yörük (2012), in a comprehensive study, empirically shows that Kurd ethnicity is the main determinant of access to social assistance. He argued that the government has adopted this strategy to prevent unrest and extend its voter base among Kurds.

In addition to the social aid provided by charity organizations, the AKP has also effectively used the conditional cash transfer program (CCT). Although this social assistance scheme was established in 1986, it was not so effective in alleviating poverty. In response to the devastating effects of the 2001 economic, the World Bank started to contribute to Turkey's social assistance scheme in 2003 through its Social Risk Mitigation Project. CCT was a core part of this project.

The goal of CCT programs is to reduce poverty in the short run and increase children's human capital endowment in the long run to prevent the intergenerational transmission of poverty. This double goal has made CCT the most popular part of social assistance programs. The program improved the well-being of children and women in Turkey. For example, food consumption of beneficiaries has increased, domestic violence has declined, and the beneficiaries have started to value education equally for girls as well as boys (Acar et al. 2012; Yildirim et al. 2014; Aydiner-Avşar

2015). However, one cannot yet claim that CCT has transformed gender relations. In fact, it is not even justifiable to argue that CCT empowers women (Emir et al. 2013; Elveren 2015a). First, the decline in domestic violence has been a temporary result of improved household financial conditions rather than a permanent change. Women are likely to face the same threat of violence, if not more, without CCT. Second, CCT can only benefit women if they have a voice in the household and the power to decide how to spend money (Yılmaz 2014). Indeed, it can be argued that CCT actually reinforces patriarchal gender relations by giving woman the responsibility for taking care of children and disabled relatives (Lomeli 2008; Acar and Altunok 2013; Bergmann and Tafolar 2014; Elveren 2015a).

5.2.2 Deep Impact: Islamization of the Education System and Imam-Hatip Schools

The evidence of the Islamization of Turkey is clearest in the education system. In addition to two ongoing fundamental problems in the education system—inequalities and low quality of teaching—the last decade has witnessed increasing encouragement to the private sector and a shift to a religious system. When the AKP came to power in 2002, it announced that addressing flaws in the education system was one of its priorities. Although subsequent reforms reduced class sizes in primary schools, they failed to address the gender gap and discrepancies between urban and rural citizens and between Turkey's West and East in terms of educational opportunities and attainment (Elveren and Ağartan 2017). The AKP government has subsidized private schools and strongly encouraged private investment in public-private partnerships. This, not surprisingly, has led to a substantial increase in citizens' out-of-pocket expenditure on education.

However, the AKP's most determined steps in education have been to increase the importance of Imam-Hatip schools, which were originally established in 1923 as vocational schools to train imams and preachers for mosques while also enabling the state to control religious education. In 1974, Imam-Hatip school graduates were permitted to choose any major, rather than just theology, subject to their performance in the university entrance examination. The number of schools increased quickly in this era until the law was changed in 1999 to make it extremely difficult for

Imam-Hatip students to choose any program except theology, resulting in a dramatic fall in student enrolments in these schools.

The AKP was able to abolish the 1999 law in 2009, and its support of Imam-Hatip schools did not stop there. In 2010, the government made it possible to transform regular high schools into Imam-Hatip schools and, in 2012, for middle schools to do this as well. Also in 2012, the highly contested 4 + 4 + 4 law increased compulsory schooling from eight to twelve years, with four years of primary education and four years of secondary education in middle schools and religious (Imam-Hatip) middle schools, thereby allowing Imam-Hatips to recruit students for their middle school sections. Accordingly, enrollment jumped from 65,000 students when AKP came to power in 2002 to about 1.3 million in the 2016–2017 academic year. The party's determined effort to reduce the number of nonreligious schools and increase the number of Imam-Hatip schools has forced some students, particularly in small towns, to go to Imam-Hatips regardless of their families' will.

The AKP's moves toward Islamization of the education system were not limited to eager encouragement of Imam-Hatip schools. The curriculum has been reformed to emphasize neoliberal and Islamic values (Türkmen 2009; Ünder 2012), for example, by excluding the theory of evolution from the school curriculum and incorporating a highly controversial concept of jihad in the Imam-Hatip curriculum. Other important changes include ending the minimum age requirements for starting Koran courses, increasing the hours for the compulsory religion class, de-emphasizing Atatürk and his principles in the curriculum, and recruiting school principals based on oral exams to ensure that all schools are run by 'pious' Muslims.

These reforms have not reduced gender gaps in access to and attainment in education, although the party had initially enthusiastically declared this as a goal. On the contrary, the 4 + 4 + 4 law has significant potential to limit female educational opportunities since the new curriculum, particularly in Imam-Hatip schools, reinforces women's subordinate position in society by reducing womanhood to motherhood and assigning women the roles of mothers and wives in the family unit.

To conclude, nothing could be more symbolic perhaps than that the AKP removed the class on 'human rights, citizenship and democracy' from the school curriculum.

5.3 Gender Inequality in Turkey[5]

5.3.1 Gender Inequality in Labor Market

The Polity IV Index 2016 categorizes Turkey's regime 'open anocracy', a regime of partial democracy and partial dictatorship. The regime has shifted further toward authoritarianism following the highly controversial referendum, which took place under a state of emergency. This radical shift in regime is likely to trigger the brain drain from Turkey 'as educated people prefer to live in more democratic and politically stable environments as well as to have freer, westernized lifestyles, individually and socially' (Elveren and Toksöz 2017, p. 18).

Turkey ranks 71st in the UN Human Development Index with a value of 0.767, which is below the average of both EU and OECD countries. Education and gender inequalities are the main areas that cause Turkey ranking poorly. Turkey ranks 110th among 188 countries in schooling rate. More specifically, while 43.5 percent of females have at least some secondary education, the same ratio is 64.8 percent for males. The Gender Inequality Index (GII) reflects gender inequalities in reproductive health, empowerment, and economic activity. In 2015, Turkey has a GII value of 0.328, ranking 69th out of 188 countries.

Women's labor force participation rates have always been low in Turkey. The labor force participation rate (employment rate) is 33.8 (29.3) percent for females but 72.1 (65.8) percent for males by November 2017. Not surprisingly, women's labor force participation rate increases along with the level of education. This is because although women with higher education who hold relatively high-paid jobs can afford hiring care workers or sending kids to private kindergartens, low educated women with low wages usually have to quit their jobs once they have children because of lack of affordable public day care (İlkkaracan 2012; Elveren and Toksöz 2017).

It is, however, important to note this increase in labor force participation rate with education level goes hand in hand with an increase in the rate of unemployment. In 2016, Turkey's unemployment rates for higher education graduates were 8.8 percent for men and as high as 16.9 percent for women. This high rate of unemployment for educated women may, therefore, encourage them to migrate (Elveren and Toksöz 2017).

Based on different surveys, several studies have shown the persistent and significant gender wage gap, around 70–77 percent in the ratio of women's earnings to that of men's earnings in the average sense (Elveren 2013b). The gap in gross national income per capita is also remarkable, $10,648 for females and $27,035 for males in 2011 PPP $. Women in Turkey face earning gaps above the global average.

There has been significant vertical and horizontal segregation by gender in the labor market in Turkey. While hierarchical segregation refers to segregation in an institution based on gender (e.g., women vs. men in top positions), vertical segregation represents segregation in different sectors based on gender (e.g., women vs. men in care/construction jobs). We note that both hierarchical and vertical segregations are major reasons behind the existing gender pay gap in Turkey (Ecevit 2008; Elveren 2013b, p. 38). For instance, the fact that only 15.1 percent of managers are women suggests that educated women face serious obstacles in their career advancement, which may influence their decision to migrate (Elveren and Toksöz 2017). Finally, women's political representation in Turkey is also low. Women hold only 14.9 percent of parliamentary seats.

Although the government in Turkey has revised the Civil Code, the Labor Law, and the Penal Code legislature toward gender equality in line with the EU initiatives, discriminatory mentalities and practices continue (Elveren 2013b; Elveren and Toksöz 2017). According to the OECD Social Institutions & Gender Index Synthesis Report (2014), although 'Turkey has a low level of discrimination regarding discriminatory family codes, restricted physical integrity, and restricted resources and assets but a high level of discrimination against daughters and restricted civil liberties' (Elveren and Toksöz 2017, p. 17).

The same report also notes that while women spend on average more than twice as much time as men in unpaid care work across OECD countries, the gap is as high as five times in Japan, Korea, and Turkey, showing a remarkable 'double burden' for women (Elveren and Toksöz 2017, p. 17). This inequality is a root cause of low female employment and a large gender income gap.

Toksöz (2016) states that AKP's female employment policies have replaced previous egalitarian rhetoric with a focus on the importance of the family for social continuity, giving women responsibility for maintaining and protecting the family. The government has proposed flexible work options for women, such as part-time work and female entrepreneurship, which result in precarious forms of employment (Toksöz 2016).

5.3.2 Education

The major problems in the education system in Turkey are inequalities between genders, socio-economic groups, and geographical regions and low quality of teaching (Elveren and Ağartan 2017). When AKP came to power in 2002, it announced addressing flaws in the education system as one of its priorities. 'Between 2002 and 2011, the Ministry of National Education implemented a curriculum reform, introduced conditional cash transfers, and carried out a campaign called "Girls, Let's Go to the School". As a result, class size at the primary levels reduced substantially, and some improvements in gender parity were seen—especially in secondary education' (Elveren and Ağartan 2017, p. 324). However, as discussed in Chap. 5, education—with social aid provision—has become the key tool for AKP to transform the regime.

There is a remarkable gender gap in education level. Among population aged 25 years and older, in 2005, there were 100 female primary school graduates for every 76 male, 100 female high school graduates for every 151 male, and 100 female higher education graduates for every 136 men.

Although there is not remarkable gender gap in higher education in terms of enrollment, there is a segregation in fields by gender. Although women are unsurprisingly concentrated in health sciences, social sciences, and art, there are 100 women for every 72 men in mathematics and natural sciences. Similarly surprising, the share of female students in engineering, manufacturing, and construction is 29.4 percent and in medicine is 49.4 percent. Moreover, as high as 30.8 percent of master's degree students in engineering, manufacturing, and construction are currently women. This has long historical roots in Turkey in the ideological approach of successive governments after the foundation of the Republic in 1923 by Mustafa Kemal Atatürk who considered that improving women's social and political status and supporting their rights in higher education and employment in professional jobs were integral to Turkey's Westernization process. Adopting a rationalist, positivist worldview, they promoted natural and technical sciences and influenced women from middle-upper-class families in their choices of these study fields (Acar 1983; Elveren and Toksöz 2017).

5.3.3 Violence Against Women

Violence against women is a serious problem in Turkey, cutting across all social groups and strata, regardless of women's level of welfare or education. Between 2010 and 2016, at least 1411 women were killed. About 53.6 percent of offenders were husbands or ex-husbands, 18.8 percent were male relatives, and 14.2 percent were boyfriends or ex-boyfriends. Perhaps even more tragic is that 234 of those murders were committed during an ongoing separation or divorce process, and 141 of them occurred although the women had applied for protection in the face of harassment or threats. This shows the inadequacy of measures taken by the public authorities to prevent violence against women.[6,7] Although the AKP government has taken legal and institutional measures to combat violence against women, their consistent implementation stays on the legislative text as the government's understanding of 'gender equality' is limited to a legal equality. In fact, in the social life, government ambitiously promotes familism and reinforces the subordinate role of women. However, 'violence against women in Turkey cannot be combatted without concrete steps towards economic, social and political gender equality' (Aslan-Akman and Tütüncü 2013; Elveren and Toksöz 2017, p. 22).

It is reasonable to argue that 'women's risk of facing violence in Turkey along with the sexist rhetoric of public authorities is likely to strengthen educated women's desire to emigrate' (Elveren and Toksöz 2017, p. 22).

Notes

1. See Akça et al. (2014) for a collection of articles on the regime change in Turkey during the AKP era.
2. http://www.hurriyetdailynews.com/erdogans-declaration-of-system-change-outrages-turkeys-opposition-.aspx?pageID=238&nID=87036&-NewsCatID=338
3. See Dedeoğlu and Elveren (2012a, b, 2015) for comprehensive analyses of the various aspects of the welfare regime in Turkey and Elveren and Ağartan (2017) for a special focus on human capital development; Elveren (2008a) and Elveren and Elveren (2010) for a critical analysis of social security reform in Turkey from a radical institutional perspective; Elveren (2015b) for an empirical analysis on the effect of women's labor force participation on social security deficits; Elveren (2013b) for a critical assessment of the pension system in Turkey from a gender equality perspective; and Elveren (2008b), Elveren and Hsu (2007), Bozkuş and Elveren (2008), Şahin and

Elveren (2011, 2014), and Şahin et al. (2010) for empirical analyses of the gender inequality in the Individual Pension System, introduced in 2003 as a part of social security reform.
4. See Bozkuş (2009) for the positive impact of the project.
5. This section benefits from Elveren and Toksöz (2017).
6. http://bianet.org/english/gender/134394-bianet-is-monitoring-male-violence
7. It is tragic (and also extremely representative of the new regime in Turkey) that a 'man was taken into custody by Turkish police the day after a complaint was made by his ex-girlfriend' for over-harassing and threating her for two years and reportedly released pending trial the following day. However, 'he was re-detained by the police the same day and this time was put in jail because police officers detected messages and e-mails on his phone that allegedly insulted President Erdoğan' https://stockholmcf.org/turkish-man-released-over-harassing-ex-girlfirend-arrested-for-insulting-president-erdogan/ (last accessed on March 13, 2018).

REFERENCES

Acar, F. (1983). Turkish Women in Academia: Roles and Careers. *METU Studies in Development*, 10, 409–446.
Acar, F., & Altunok, G. (2013). The 'politics of intimate' at the intersection of neo-liberalism and neo-conservatism in contemporary Turkey. *Women's Studies International Forum*, 41, 14–23.
Acar, M., Köse, N., Yıldırım Ocal, J., Boyacı, A., & Sezgin, F. (2012). *Qualitative and Quantitative Analysis of Impact of Conditional Cash Transfer Program in Turkey Project Report*. Ankara: Ministry of Family and Social Policies.
Akça, İ., Bekmen, A., & Özden, B. A. (Eds.) (2014). *Turkey Reframed Constituting Neoliberal Hegemony*. London: Pluto Press.
Aslan-Akman C., & Tütüncü, F. (2013). The Struggle against Male Violence with an Egalitarian Jurisprudence and Religious Conservative Government: The Case of Secular Turkey. In Z. S. Salhi (Ed.), *Gender and Violence in Islamic Societies* (pp. 82–107). London: I. B. Tauris.
Atalay, Z. (2017). Partners in Patriarchy: Faith-Based Organizations and Neoliberalism in Turkey. *Critical Sociology*. https://doi.org/10.1177/0896920517711488
Aydin, E., Hisarciklilar, M., & Ilkkaracan, I. (2010). Formal versus informal labor Market Segmentation in Turkey in the Course of Market Liberalization. *Topics in Middle Eastern and North African Economies*, 12, 1–43.
Aydıner-Avşar, N. (2011). *Essays on Trade and Wage Structure in Turkey*. Unpublished PhD dissertation, Department of Economics, The University of Utah, Salt Lake City, Utah, USA.

Aydıner-Avşar, N. (2015). Conditional Cash Transfer Programs from a Gender Perspective: A Comparative Evaluation for Turkey. *European Journal of Economic and Political Studies*, 7, 37–66.

Bedford, K. (2008). Holding it together in a crisis: Family strengthening and embedding neoliberalism. *IDS Bulletin*, 39(6), 60–66.

Bergmann, C., & Tafolar, M. (2014). Combating Social Inequalities in Turkey through Conditional Cash Transfers (CCT)? Conference Proceeding, the 9th Global Labour University Conference "Inequality within and among Nations: Causes, Effects, and Responses", Berlin School of Economics and Law.

Boratav, K. (1990). Inter-Class and Intra-Class Relations of Distribution Under 'Structural Adjustment': Turkey During the 1980s. In T. Aricanli & D. Rodrik (Eds.), *The Political Economy of Turkey: Debt, Adjustment and Sustainability* (pp. 199–229). London: Macmillan.

Boratav, K., Yeldan, E., & Köse, A. H. (2002). Globalization, distribution and social policy: Turkey: 1980–1998. In L. Taylor (Ed.), *External Liberalization and Social Policy*. London: Oxford University Press.

Bozkuş, S. C. (2009). *Importance of Human Capital and Infrastructure for Turkish Regions*. Unpublished Master's Thesis, Department of Sociology, The University of Utah, Salt Lake City, Utah, USA.

Bozkuş, S. C., & Elveren, A. Y. (2008). An Analysis of Gender Gaps in the Private Pension Scheme in Turkey. *Ekonomik Yaklaşım*, 19(69), 89–106.

Buğra, A. (2012). The changing welfare regime of Turkey: Neoliberalism, cultural conservatism and social solidarity redefined. In S. Dedeoğlu & A. Y. Elveren (Eds.), *Gender and society in Turkey: The impact of neo-liberal policies, political Islam and EU accession* (pp. 15–31). London and New York: I.B. Tauris.

Buğra, A. (2013). Revisiting the Wollstonecraft dilemma in the context of conservative liberalism: The case of female employment in Turkey. *Social Politics*, 21(1), 148–166.

Buğra, A., & Candaş, A. (2011). Change and Continuity under an Eclectic Social Security Regime: The Case of Turkey. *Middle Eastern Studies*, 47(3), 515–528.

Buğra, A., & Keyder, Ç. (2006). The Turkish Welfare Regime in Transformation. *Journal of European Social Policy*, 16(3), 211–228.

Çağatay, N. (1986). *Inter-Industry Structure of Wages and Markups in Turkish Manufacturing*. Unpublished PhD dissertation, Department of Economics, Stanford University, CA, USA.

Dedeoğlu, S. (2013). Veiled Europeanisation of Welfare State in Turkey: Gender and Social Policy in the 2000s. *Women's Studies International Forum*, 41(1), 7–13.

Dedeoğlu, S., & Elveren, A. Y. (Eds.) (2012a). *Gender and Society in Turkey: The Impact of Neo-Liberal Policies, Political Islam and EU accession*. London & New York: I.B. Tauris Publishers.

Dedeoğlu, S., & Elveren, A. Y. (Eds.) (2012b). *Türkiye'de Refah Devleti ve Kadın [The Welfare State and Woman in Turkey]*. İstanbul: İletişim Yayınevi.

Dedeoğlu, S., & Elveren, A. Y. (Eds.) (2015). *2000'ler Türkiye'sinde Sosyal Politika ve Toplumsal Cinsiyet [Social Policy and Gender in Turkey in the 2000s]*. Ankara: İmge Kitabevi Yayınları.
Derin-Güre, P., & Elveren, A. Y. (2014). Does Income Inequality Derive Separatist Terrorism in Turkey? *Defence and Peace Economics*, 25(3), 311–327.
Ecevit, Y. (2008). İşgücüne Katılım ve İstihdam (Labour Force Participation and Employment). In M. Tan, S. Sancar & S. Acuner (Eds.), *Türkiye'de Toplumsal Cinsiyet Eşitsizliği: Sorunlar, Öncelikler ve Çözüm Öneriler* (pp. 113–213). İstanbul: İstanbul TÜSİAD Publications.
Elgin, C., & Kuzubaş, T. U. (2012). Wage-Productivity Gap in Turkish Manufacturing Sector. *Iktisat Isletme ve Finans*, 27(316), 9–31.
Elveren, A. Y. (2008a). Social Security Reform in Turkey: A Critical Perspective. *Review of Radical Political Economics*, 40(2), 212–232.
Elveren, A. Y. (2008b). Assessing Gender Inequality in the Turkish Pension System. *International Social Security Review*, 61(2), 39–58.
Elveren, A. Y. (2010). Wage Inequality in Turkey: Decomposition by Statistical Regions, 1980–2001. *Review of Urban & Regional Development Studies*, 22(1), 55–72.
Elveren, A. Y. (2012). Military Spending and Income Inequality: Evidence on Cointegration and Causality for Turkey, 1963–2007. *Defence and Peace Economics*, 23(3), 289–301.
Elveren, A. Y. (2013a). A Brief Note on Deunionization and Pay Inequality in Turkey. The University of Texas Inequality Project Working Paper, No. 63.
Elveren, A. Y. (2013b). A critical analysis of the pension system in Turkey from a gender equality perspective. *Women's Studies International Forum*, 41(1), 35–44.
Elveren, A. Y. (2015a). Türkiye'de Sosyal Güvenlik Sisteminin Toplumsal Cinsiyet Eşitliği Açısından Bir Değerlendirmesi. In S. Dedeoğlu & A. Y. Elveren (Eds.), *2000'ler Türkiye'sinde Sosyal Politika ve Toplumsal Cinsiyet* (pp. 63–91). Ankara: Imge Yayinevi.
Elveren, A. Y. (2015b). The Impact of the Informal Employment on the Social Security Deficits in Turkey. *World Journal of Applied Economics*, 1(1), 3–19.
Elveren, A. Y. (2018). The Pious Predator State: The New Regime in Turkey. *Challenge*, 61(1), 85–91.
Elveren, A. Y., & Ağartan, T. İ. (2017). The Turkish Welfare State System: With Special Reference to Human Capital Development. In C. Aspalter (Ed.), *The Routledge International Handbook to Welfare State Systems*. New York and London: Routledge.
Elveren, M. A., & Elveren, A. Y. (2010). The Transformation of the Welfare Regime in Turkey and the Individual Pension System. *Mülkiye*, 34(266), 243–258.
Elveren, A. Y., & Galbraith, J. K. (2009). Pay Inequality in Turkey in the Neo-Liberal Era, 1980–2001. *European Journal of Comparative Economics*, 6(2), 177–206.

Elveren, A. Y., & Hsu, S. (2007). *Gender Gaps in the Individual Pension System in Turkey*. The University of Utah, Department of Economics, Working Paper No. 6.

Elveren, A. Y., Ornek, I., & Akel, G. (2012). Internationalisation, growth and pay inequality: A cointegration analysis for Turkey, 1970–2007. *International Review of Applied Economics*, 26(5), 579–595.

Elveren, A. Y., & Özgür, G. (2016). The Effect of Informal Economy on Income Inequality: Evidence from Turkey. *Panoeconomicus*, 63(3), 293–312.

Elveren, A. Y., & Toksöz, G. (2017). Why Don't Highly Skilled Women Want to Return? Turkey's Brain Drain from a Gender Perspective. MPRA No. 80290, 2017. https://mpra.ub.uni-muenchen.de/80290/

Emir, İ., Erbaydır, T., & Yüksel, A. (2013). Şartlı Nakit Transferi Uygulaması Kadınların Toplumsal Konumunu Değiştiriyor mu? *Fe Dergi: Feminist Eleştiri*, 5(2), 120–133.

Eres, B. (2005). *The Profit Rate in the Turkish Economy: 1968–2000*. Unpublished PhD dissertation, Department of Economics, The University of Utah, Salt Lake City, Utah, USA.

Eres, B. (2011). Adalet ve Kalkınma Partisi'nin Siyasî Başarısının Kaynakları Üzerine Kısa bir Not. *Mülkiye*, 35(271), 163–174.

Galbraith, J. K. (2008). *The Predator State How Conservatives Abandoned the Free Market and Why Liberals Should Too*. New York: Free Press.

Heper, M. (1997). Islam and Democracy in Turkey: Toward a Reconciliation? *Middle East Journal*, 51(1), 32–45.

İçduygu, A., Romano, D., & Sirkeci, I. (2010). The ethnic question in an environment of insecurity: The Kurds in Turkey. *Ethnic and Racial Studies*, 22(6), 991–1010.

İlkkaracan, İ. (2012). Why there are so few women in the labor market in Turkey? A multi dimensional approach. *Feminist Economics*, 18(1), 1–37.

Karadeniz, O. (2012). *Asisp Annual Report 2012 Turkey: Pensions, Health Care and Long Term Care*. http://pensionreform.ru/files/13660/ASISP.%-20Annual%20National%20Report%202012%20-%20Turkey.pdf

Kızılırmak Yakışır, A. B. (2003). *Explaining Wage Inequality: Evidence from Turkey*. Ankara University Faculty of Political Science Research Center for Development and Society Working Paper 57.

Larner, W. (2000). Post-welfare state governance: Towards a code of social and family responsibility. *Social Politics: International Studies in Gender, State & Society*, 7(2), 244–265.

Lomeli, E. V. (2008). Conditional Cash Transfers as Social Policy in Latin America: An Assessment of their Contributions and Limitations. *The Annual Review of Sociology*, 34, 475–499.

Memis, E. (2007a). *Inter- and Intraclass Distribution of Income in Turkish Manufacturing, 1970–2000*. Unpublished PhD dissertation, Department of Economics, The University of Utah, Salt Lake City, Utah, USA.

Memis, E. (2007b). A Disaggregate Analysis of Profit Rates in Turkish Manufacturing. *Review of Radical Political Economics*, 39(3), 398–406.

Meschi, E., Taymaz, E., & Vivarelli, M. (2008). *Trade Openness and the Demand for Skills: Evidence from Turkish Microdata*. Institute for the Study of Labor Discussion Paper 3887.

Onaran, Ö. (2009). Wage Share, Globalization and Crisis: The Case of the Manufacturing Industry in Korea, Mexico and Turkey. *International Review of Applied Economics*, 23(2), 113–134.

Ongan, T. H. (2011). Profit Rate of Turkish Manufacturing Sector in a Marxian Perspective. *Sosyal Bilimler Dergisi*, 1, 1–10.

Öniş, Z. (1997). The Political Economy of Islamic Resurgence in Turkey: The Rise of the Welfare Party in Perspective. *Third World Quarterly*, 18(4), 743–766.

Oyvat, C. (2011). Globalization, Wage Shares and Income Distribution in Turkey. *Cambridge Journal of Regions, Economy and Society*, 4(1), 123–138.

Özden, B. A., Akça, İ., & Bekmen, A. (2017). Antinomies of Authoritarian Neoliberalism in Turkey: The Justice and Development Party Era, In C. B. Tansel (Ed.), *States of Discipline: Authoritarian Neoliberalism and the Contested Reproduction of Capitalist Order* (pp. 189–209). London and New York: Rowman & Littlefield International.

Polanyi, K. 1944 [2001]. *The Great Transformation The Political and Economic Origins of Our Time*. Boston: Beacon Press.

Şahin, Ş., & Elveren, A. Y. (2011). Assessing a Minimum Pension Guarantee for the voluntary IPS in Turkey. *International Social Security Review*, 64(3), 39–61.

Şahin, Ş., & Elveren, A. Y. (2014). A Minimum Pension Guarantee Application for Turkey: A Gendered Perspective. *Journal of Women, Politics & Policy*, 35(3), 242–270.

Şahin, Ş., Rittersberger-Tılıç, H., & Elveren, A. Y. (2010). The Individual Pension System in Turkey: A Gendered Perspective. *Ekonomik Yaklaşım*, 21(77), 115–142.

Smith, A. M. (2008). Neoliberalism, welfare policy, and feminist theories of social justice. *Feminist Theory*, 9(2), 131–144.

Toksöz, G. (2016). Transition from 'Woman' to 'Family', An Analysis of AKP Era Employment Policies from a Gender Perspective. *Journal für Entwicklungspolitik*, 32(1/2), 64–83.

Türkmen, B. (2009). A transformed Kemalist Islam or a new Islamic civic morality? A study of "religious culture and morality" textbooks in the Turkish high school curricula. *Comparative Studies of South Asia, Africa and the Middle East*, 29(3), 381–397.

Ünder, H. (2012). Constructivism and the Curriculum Reform of the AKP. In K. İnal & G. Akkaymak (Eds.), *Neoliberal Transformation of Education in Turkey Political and Ideological Analysis of Educational Reforms in the Age of the AKP* (pp. 33–45). New York: Palgrave Macmillan.

Voyvoda, E., & Yeldan, E. (2001). Patterns of Productivity Growth and the Wage Cycle in Turkish Manufacturing. *International Review of Applied Economics*, 15(4), 375–396.

Yeldan, E. (2006). Neoliberal Global Remedies: From Speculative-Led Growth to IMF-Led Crisis in Turkey. *Review of Radical Political Economics*, 38(2), 193–213.

Yildirim, J., Ozdemir, S., & Sezgin, F. (2014). A Qualitative Evaluation of a Conditional Cash Transfer Program in Turkey: The Beneficiaries' and Key Informants' Perspectives. *Journal of Social Service Research*, 40(1), 62–79.

Yılmaz, B. (2014). Türkiye'de Sosyal Devletin Dönüşümü ve Sosyal Yardımlar: Şartlı Nakit Transferi Alan Kadınlar Üzerine Bir Değerlendirme. In K. Akkoyunlu Ertan, F. Kartal & Y. Şanlı Atay (Eds.), *Sosyal Adelet için İnsan Hakları: Sosyal Haklar Bildiriler Kitabı*. Ankara: TODAİE.

Yörük, E. (2012). Welfare Provision as Political Containment: The Politics of Social Assistance and the Kurdish Conflict in Turkey. *Politics & Society*, 40(4), 517–547.

Yörük, E., & Özsoy, H. (2013). Shifting forms of Turkish state paternalism toward the Kurds: Social assistance as "benevolent" control. *Dialectical Anthropology*, 37(1), 153–158.

CHAPTER 6

Conclusion and Policy Recommendations

Abstract This chapter summarizes the main findings of the analysis on the return intentions of the Turkish students and professionals residing abroad, with a special focus on the gender aspect. Then, the chapter discusses the challenges Turkey faces and provides policy recommendations regarding brain drain.

Keywords Brain drain • Gender • High-skilled migration • Push-pull models

The purpose of this work was to examine the brain drain in Turkey. The study examined the return intentions of the students and professionals, with a special focus on the gender dimension. Gender dimension of the brain drain has not received enough attention, not just in Turkey but also in international studies. The study is based on data of 116 students and 84 professionals derived from two comprehensive questionnaires, which include 46 questions and 53 questions in the case of professionals and students, respectively.

Brain drain is the migration of highly educated individuals from their home countries to developed countries in order to reach greater opportunities in their field of specialization and/or to have better living conditions and lifestyle. It is an important item in the agenda of policymakers in developing countries as it is considered one of the most detrimental aspects of international migration from the viewpoint of home country.

© The Author(s) 2018
A. Y. Elveren, *Brain Drain and Gender Inequality in Turkey*,
https://doi.org/10.1007/978-3-319-90860-1_6

Brain drain is the 'reverse technology transfer', a cheap means for developed countries to acquire high-skilled labor. It has substantial costs for the sending country because in the absence of such a group of highly educated individuals, the information and knowledge cannot transmit in the society. Moreover, there will be less people to demand for better governance and civic society, which together reduce the productivity in the economy. Another cost of brain drain for the sending country is that they experience difficulty in the delivery of health and education services at desired levels. This deteriorates the long-term growth potential of the country because the key element of economic growth is human capital.

Some scholars, on the other hand, argue that the brain drain can benefit the sending country thanks to 'skill mobility'. The issue of brain drain is highly crucial for Turkey as it has long been one of the top ten countries that send their students to pursue graduate degrees in the US, the top destination for foreign students to pursue graduate degrees. It is important to understand the causes of brain drain and take some measures to prevent it in order to get maximum benefit from these highly skilled individuals. Therefore, this research is a modest attempt to provide some more empirical evidence on the subject matter to help the policymakers to set up the long-term development strategy.

Despite sizeable literature on migration in Turkey, there are few studies that consider the brain drain. In fact, there are only a few comprehensive empirical investigations. This study contributes to the Turkish literature by providing some fresh evidence. The study supports most of the previous findings of the literature and provides some updated and different information. Accordingly, the effectiveness of compulsory service rule; insignificance of the economic condition in Turkey on the return decision; the political instability being the key factor on return decision, the positive relationships between stay time abroad, having degree from a foreign institution/or a university whose medium of language is English, and previous work/travel/study experience abroad and tendency not to return; no effect of the economic crisis in the US on the return decision; and higher tendency to migrate/stay abroad for females are supported by this study as well.

The study confirmed that aforementioned findings are valid today, and some of them have become more significant. For instance, political instability as a 'push' factor becomes more dominant compared to other push (and pull) factors. Also, as a result of the changing paradigm toward the role of women in the society (increasing pressure on women in social sphere and trying to confine women in the private sphere), women have a higher tendency than men to migrate and not to return. The paper also found that female students are more likely to receive support from their families to go abroad. Moreover, it is found that women have higher tendency to go abroad and prefer to stay more because of push factors than because of pull factors.

What should be done? First of all, not just this study but also some other major works showed that the most significant concern of students/academics is the 'political instability' in general and 'the lack of academic freedom' in particular. In fact, as some studies highlighted, students/academics consider the wage difference as a secondary issue in return decision. In other words, it is safe to argue that the financial incentives will have limited effect on the reverse brain drain as long as social polarization and the pressure on the academia continue. Therefore, the key issue is to create a 'suitable' environment for academicians to pursue their research. In fact, the policymakers should remember the robust link between democracy and economic development as emphasized, among many others, by the leading Turkish economist Daron Acemoğlu.

Second, it is a fact that the initiatives by TÜBİTAK in terms of the reverse brain drain were successful to a certain degree. Perhaps, the basic shortcoming of the campaign is that policymakers think that the financial incentives are sufficient to retain scholars abroad. In this context, the valuable insights of scholars on the initiatives of the TÜBİTAK provided in Esen (2014) should be considered carefully. The majority of the scholars view this initiative problematic in terms of its applicability. They also emphasize that it lacks adequate research equipment and research opportunities. Also, they argue that it is 'mainly for hard science professors, and monetary rather than research and development focused' (ibid., p. 54). That is, the initiatives should be designed with respect to insights by scholars.

Third, although there has been significant increase in the funds devoted to the research and development (R&D), they are still below the average of advanced countries. Similarly, despite the substantial increase in expenditure on all levels of educational institutions, Turkey ranks the lowest in OECD in terms of the cumulative expenditure per student over the average duration of tertiary studies (OECD 2014, cited in Elveren and Ağartan 2017).

It is important not just to devote more funds to the R&D but also to increase the efficiency of the use of these funds. Increasing concerns about some practices of TÜBİTAK, widening gap between the quality and quantity of works produced in the universities (mostly because of newly established universities), and the overall decline in the quality of education should definitely be considered by the policymakers. The number of universities jumped from 70 in 2003 to 196 in 2014, with substantial deficit in the number of 'competent' instructors and infrastructure. It is an unfortunate fact that the majority of these universities have very poor research performance (Tekneci 2014).

Moreover, 'discrepancies in educational opportunities and attainment between urban and rural citizens, genders, and the West and East regions of the country' are serious challenges for Turkey (World Bank 2006, cited in Elveren and Ağartan 2017). It is a fact that increase in expenditure on educational institutions has not led to clear improvements in basic educational indicators. Compared to OECD countries, Turkey ranks either the lowest or second lowest after Mexico with respect to many basic educational indicators (Elveren and Ağartan 2017). Turkey has a higher-than-average proportion of underperforming students (OECD 2014, cited in Elveren and Ağartan 2017). Students from Turkey do not perform well in the OECD Program for International Student Assessment.

Regarding the major reform introduced in 2012, Elveren and Ağartan (2017, p. 326) note that

> [s]ome critics highlighted the shift to a value based education system where the teaching of religion and religious middle schools were introduced to the mainstream education system. Others drew attention to the negative consequences of this new 4 + 4 + 4 system on gender equality because of the law's potential for limiting educational opportunities for females (Kader 2012). Similar to the health care reform, major social actors are excluded from policymaking, and the views of main providers of education—the teachers—are not seriously incorporated in the policy debate.

Fourth, the prestige associated with having a foreign degree is substantial. Therefore, it is inevitable that most of the qualified students might consider the opportunity to study abroad. On the other hand, major advances in communications technology reduce the extent to which skills are actually lost. Therefore, in this context, it is important to focus on 'brain circulation' rather than 'brain drain'. In other words, the policymakers should emphasize the policies to increase the possible linkages between scholars in Turkey and the Turkish scholars abroad. Increasing the number of joint research institutions and joint programs (such as dual diploma programs) and introducing novel incentives to encourage collaboration between scholars in Turkey and abroad should be considered by the policymakers.

Finally, the link between students' study fields abroad and demand by industry should be tightened. Also, the mismatch between available jobs, the skills required in the market, and the courses available in the higher education system must be eliminated. Moreover, making a clear separation of research and teaching universities as has been implemented abroad and increasing the autonomy of the universities may help to achieve this goal.

REFERENCES

Elveren, A. Y., & Ağartan, T. İ. (2017). The Turkish Welfare State System: With Special Reference to Human Capital Development. In C. Aspalter (Ed.), *The Routledge International Handbook to Welfare State Systems*. New York and London: Routledge.

Esen, E. (2014). *Going and Coming: Why U.S.-Educated Turkish PhD Holders Stay in the U.S. or Return to Turkey?* Unpublished PhD dissertation, Department of Educational Leadership and Policy Studies, The University of Kansas, Lawrence, KS, USA.

Kader. (2012). *Question from Kader to President Abdullah Gul about 4 + 4 + 4.* http://www.ucansupurge.org/turkce/index2.php?Hbr=584. Accessed 11 July 2016.

OECD. (2014). *Education at a Glance 2014: OECD Indicators*. OECD Publishing.

Tekneci, P. D. (2014). *Evaluating Research Performance of Turkish Universities*. Unpublished PhD Dissertation, Middle East Technical University, Ankara, Turkey.

World Bank. (2006). *Turkey Public Expenditure Review*. Report No. 36764-TR. Washington, DC: The World Bank.

Appendix

Brain Drain Professional Survey

Dear friend/colleague,

I am an assistant professor of economics at Fitchburg State University, MA. I have been conducting a funded research to analyze the brain drain, a very crucial issue for Turkey.

With your invaluable contribution I can comprehensively analyze the decision that Turkish students and professionals make in going abroad and returning (or not returning), obtaining important results both at the theoretical and policy-making levels.

The survey involves—in addition to information on your age, gender, and hometown—several questions on your educational background and work experience as well as on your decision to stay abroad or return to Turkey.

This survey minimizes risk for participants. It does not collect name, email address, or any sort of contact information. Data will be used in a de-identified way. The information you will be providing will be used solely for statistical/econometric analysis for a scholarly article, and the information is not going to be shared with third parties.

Data will be stored in a password-protected personal computer for three years.

The survey will take about 10–15 minutes. You will be able to skip questions that are not applicable in your case. In return, your inputs will help to better understand an important issue, brain drain, for developing countries, particularly for the case of Turkey.

You may ask any questions or make suggestions about the survey by emailing me at aelveren@fitchburgstate.edu. My mailing address is 160 Pearl Street, Miller Hall 307, Fitchburg, MA, 01420, USA, and my phone number is +1-978-665-4855.

Regarding anonymity and confidentiality or any other issues about the survey, you may also contact the Fitchburg State University Human Subjects Committee at humansubjects@fitchburgstate.edu.

I would be so happy if you let your Turkish friends know about this survey.

If you are enrolling in a foreign language program, are a student in a bachelor's, a master's, or a doctoral program, or a postdoctoral research fellow in the US, England, Canada, Australia, or New Zealand, please complete the Survey for Students.

If you are holding at least a bachelor's degree and working (academic or nonacademic) in the US, England, Canada, Australia, or New Zealand, please complete the Survey for Professionals.

GENERAL INFORMATION

Please read and answer carefully. The survey will take about 15 minutes. You will be able skip questions that are not applicable in your case.

1. Personal Information: Please indicate your
a) Gender:
[ ] Male
[ ] Female

2b) Birth year: _____

c) Birthplace:
city: _____
country: _____

3. a) What is your current *country* of residence?
[ ] Australia
[ ] Canada

[ ] England
[ ] New Zealand
[ ] USA
[ ] other, please indicate: _____

4. How long have you been living in your current COUNTRY of residence?
Years_____ Months_____

5.) Did you have any study, work, travel or other experience outside Turkey prior to coming to your current country of residence?
[ ] yes
[ ] no

6) **What kind of previous experience did you have abroad?**
Please mark all that apply.
[ ] study
[ ] work
[ ] travel
[ ] other, please specify: _____

7) **What is the longest period you have spent outside of Turkey, <u>not including</u> your current stay abroad?**
Years_____Months

EDUCATIONAL INFORMATION

8) **From which university did you earn your undergraduate (bachelor's or associate's) degree?**
_____

9) **What was your major? (Please indicate both if you have double majors)**
_____

10) **In Which year did you graduate?** _____

11) **What is the highest academic degree you hold?**
[ ] associate's
[ ] bachelor's
[ ] post baccalaureate certificate
[ ] master's
[ ] post master's certificate
[ ] doctorate

12) In which country did you earn your highest academic degree?

[ ] Turkey
[ ] Australia
[ ] Canada
[ ] England
[ ] New Zealand
[ ] USA
[ ] Other, please indicate: _____

13) What was your field of study?

Please be specific and indicate any areas of specialization as well.

General field of study: _____
Specialization 1: _____
Specialization 2: _____
Specialization 3: _____

14) If you received your last academic degree outside Turkey, where did you start your first full time job after completing your studies?

[ ] same city and country where I received my last degree
[ ] same country but different city
[ ] Turkey
[ ] another country, please specify: _____
[ ] not applicable

15) What is the highest academic title you hold or have held in the past?

a) in Turkey:

[ ] none
[ ] professor
[ ] associate professor
[ ] assistant professor
[ ] instructor / lecturer
[ ] research assistant
[ ] teaching assistant

b) in your current country of residence:
[ ] none
[ ] professor
[ ] associate professor
[ ] assistant professor
[ ] instructor / lecturer
[ ] research assistant
[ ] teaching assistant

WORK-RELATED INFORMATION

16) What is your occupation?
Please be specific. For example, university professor in economics, computer programmer, mid-level manager etc.

_____

17) What is your current employment status?
[ ] self-employed
[ ] employee
[ ] unemployed, looking for a job

If you are unemployed or between jobs, please refer to your last job when answering questions concerning your 'current workplace or institution'.

18 a) How long have you been working outside Turkey?
_____ number of years

b) How long have you been working in your current country of residence?
_____ number of years

c) How long have you been working at your current workplace/institution?
_____ number of years

19. How many different organizations have you worked for full time so far?
in Turkey: _____
abroad: _____

20) What sector is the firm / organization you are currently working for in?
[ ] private
[ ] public
[ ] other (e.g., non-profit organization or trust)

21) When was the firm or organization you are working for established?
[ ] within the past year (Jan. 1 2015 – Dec. 31, 2015)
[ ] within the last 2 years
[ ] within the last 5 years
[ ] within the last 10 years
[ ] 10-30 years ago
[ ] more than 30 years ago
[ ] don't know

22) Approximately how many people currently work full time in your organization (at all levels)?
[ ] less than 5
[ ] 5-11
[ ] 11-25
[ ] 26-50
[ ] 51-100
[ ] 101-200
[ ] 201-500
[ ] 501-1000
[ ] more than 1000
[ ] don't know

23) In which country were you residing when you found (or established) your current job abroad?

[ ] in my current country of residence

[ ] in Turkey

[ ] in a third country, please specify: _____

24 a) Through which channel(s) did you find your current job?

*Please mark **all** that apply.*

[ ] Direct contacts initiated with firm / organization (e.g., sending unsolicited CV)

[ ] Professional recruiters (e.g., "headhunters")

[ ] 'Career Days' held at Turkish universities

[ ] Informal channels (e.g., friends, colleagues)

[ ] Ads in professional journals

[ ] Turkish internet network (e.g., alumni networks)

[ ] Newspaper ads

[ ] Placement office at university

[ ] Faculty or advisors

[ ] other, please specify: _____

b) How did you find your first full time job abroad?

*Please mark **all** that apply.*

[ ] Direct contacts initiated with firm / organization (e.g., sending unsolicited CV)

[ ] Professional recruiters (e.g., "headhunters")

[ ] 'Career Days' held at Turkish universities

[ ] Informal channels (e.g., friends, colleagues)

[ ] Ads in professional journals

[ ] Turkish internet network (e.g., alumni networks)

[ ] Newspaper ads

[ ] Placement office at university

[ ] Faculty or advisors

[ ] other, please specify: _____

QUESTIONS RELATING TO THE DECISIONS TO LEAVE, STAY AND RETURN

25 a) What were your main reasons for going to the country you are currently residing in?

*Please mark **all** that apply.*

[ ] A. To learn a new language / improve language skills
[ ] B. In need of change / want to experience a new culture
[ ] C. Education or experience in another country is required by employers in Turkey
[ ] D. Could not find a job in Turkey
[ ] E. No program in my specialization in Turkey
[ ] F. Insufficient facilities, lack of necessary equipment to carry out research in Turkey
[ ] G. In order to take advantage of the prestige and advantages associated with study abroad
[ ] H. Preference for the lifestyle in my current country of residence.
[ ] I. To be with spouse or loved one
[ ] J. To provide a better environment for children
[ ] K. To get away from the political environment in Turkey
[ ] L. other, please specify: _____

26) Which of the above was the most important reason?

_____

27) In general, how supportive was your family (e. g. father, mother, spouse) in your decision to go abroad to work or study?

[ ] very supportive
[ ] supportive
[ ] moderately supportive
[ ] not at all supportive
[ ] not applicable

28) Before you left Turkey, what were your thoughts about returning?
[ ] I thought that I would definitely return.
[ ] I was undecided about returning; I would wait and see.
[ ] I did not think that I would return.

29) What are your thoughts about returning to Turkey now?
[ ] I will definitely return and have made plans to do so.
[ ] I will definitely return but have not made concrete plans to do so.
[ ] I will probably return.
[ ] I don't think that I will be returning.
[ ] I will definitely not return.

If you marked one of the last two options ('not return') please **question 32.**

30) What are your main reasons for returning to Turkey?
Please mark all that apply.
[ ] to complete compulsory military service
[ ] to complete university service (e.g., MEB, YÖK, TÜBITAK scholarship recipients)
[ ] I will return when my permitted time for working abroad ends (e.g. I am a visiting scholar)
[ ] I miss my family in Turkey
[ ] I want my children to continue their education in Turkey
[ ] after achieving specific goals (gaining work experience, completing research project) I want to apply what I have learned in Turkey
[ ] I will return after reaching my savings goal
[ ] I will return after reaching my career goal
[ ] I received a job offer from a firm or institution in Turkey

[ ] I want to spend my retirement in Turkey.
[ ] I don't feel safe in my current environment
[ ] other, please specify: _____

31) After you return, do you plan to go abroad again?
[ ] No
[ ] Yes
[ ] Maybe

32) In general, how does your life in your current country of residence compare with your life in Turkey?

a) work environment (e.g. your job satisfaction):
[ ] much better
[ ] better
[ ] neither better or worse
[ ] worse
[ ] much worse
[ ] not applicable

b) social aspects (e.g. friendships, social relations):
[ ] much better
[ ] better
[ ] neither better or worse
[ ] worse
[ ] much worse

c) standard of living:
[ ] much better
[ ] better
[ ] neither better or worse
[ ] worse
[ ] much worse

33 a) What are the main difficulties that you have faced / are facing living in your current country of residence? *Please mark all that apply.*

[ ] A. Being away from family
[ ] B. Children growing up in a different culture
[ ] C. Loneliness, not being able to adjust
[ ] D. Fast-paced life
[ ] E. Little or no leisure time
[ ] F. Unemployment
[ ] G. No jobs in my area of specialty
[ ] H. Discrimination against foreigners
[ ] I. Lower income compared to the income I had in Turkey
[ ] J. Higher taxes
[ ] K. Crime, lack of personal security
[ ] L. High cost of living
[ ] M. Other, please specify: _____

34) Which of the above factors do you consider to be the most difficult for you?
_____

35) Which of the follow ing factors were important in helping you adjust to life abroad? *Please mark all that apply.*

[ ] A. having previous experience abroad
[ ] B. the passage of time
[ ] C. support from the Turkish Student Association (TSA) at my institution
[ ] D. having spouse or other loved one with me
[ ] E. having cultural attaché / embassy support
[ ] F. having Turkish friends/colleagues at my university/college/research center
[ ] G. existence of a large Turkish community in my city
[ ] H. being able to share experie nces, ask for advise via Turkish internet network
[ ] I. other, please specify: _____

36) Which has been the most important factor in helping you adjust?

_____

37) What are the grea test difficulties RELATING TO TURKEY that may cause you NOT to return? Please indicate how important for you the following factors are in this decision.

Please answer even if you have indicated that you will definitely return.

REASON Very Somewhat Not Not at all
Important Important Important Important Important
    5   4   3   2   1

A. Low income in my occupation ___ ___ ___ ___ ___
B. Little opportunity for advancement ___ ___ ___ ___ ___
in my occupation
C. Limited job opportunities in my field of ___ ___ ___ ___ ___
expertise
D. No opportunity for advanced ___ ___ ___ ___ ___
training in my field
E. Being far from important research ___ ___ ___ ___ ___
centers and as a result from new
advances
F. Lack of financial resources and ___ ___ ___ ___ ___
opportunities to start up my business
G. Less than satisfying social and ___ ___ ___ ___ ___
cultural life
H. Bureaucracy, inefficiencies in ___ ___ ___ ___ ___
organizations
I. Political pressures, discord ___ ___ ___ ___ ___
J. Lack of social security ___ ___ ___ ___ ___
K. Economic instability, uncertainty ___ ___ ___ ___ ___
L. Other reason, please indicate below: ___ ___ ___ ___ ___

_____

38) Please indicate the relative importance FOR YOU of each of the following factors relating to your CURRENT COUNTRY OF RESIDENCE in deciding not to return or postpone returning to Turkey.

Please answer even if you have indicated that you will definitely return.

REASON Very Somewhat Not Not at all
Important Important Important Important Important
     5   4   3   2   1

A. Higher salary or wage ___ ___ ___ ___ ___

B. Greater opportunity to advance ___ ___ ___ ___ ___
in profession

C. Better work environment ___ ___ ___ ___ ___
(flexible work hours, relaxed setting, etc.)

D. Greater job availability in my area ___ ___ ___ ___ ___
of specialization

E. Greater opportunity for further ___ ___ ___ ___ ___
development in area of specialty

F. A more organized and ordered life ___ ___ ___ ___ ___
in general

G. More satisfying social and cultural life ___ ___ ___ ___ ___

H. Proximity to important research ___ ___ ___ ___ ___
and innovation centers

I. Spouse's preference to stay or ___ ___ ___ ___ ___
spouse's job being in current country

J. Better educational opportunities for children / ___ ___ ___ ___ ___
want children to continue their education

K. Need to finish or continue with current project ___ ___ ___ ___ ___

L. Other reason, please specify below: ___ ___ ___ ___ ___

_____

APPENDIX 151

Other information

39 a) Are you a member of any professional, cultural or alumni associations / societies?
[ ] Yes
[ ] No

40) What type of positive contribution(s) do you think your stay abroad is making or has made to Turkey? *Please mark all that apply.*
[ ] Helped Turkish students find scholarships abroad
[ ] Participated in lobbying activities on behalf of Turkey
[ ] Helped increase business contacts with Turkey
[ ] Helped increase knowledge about Turkey in general
[ ] Made donations to Turkish organizations
[ ] Helped increase professional contacts between colleagues in my current country and colleagues in Turkey
[ ] Helped transfer knowledge gained in my current country of residence to colleagues in Turkey (e.g., by presenting papers in conferences or teaching in Turkey)
[ ] other, please specify: _____

41) Please indicate your marital status:
[ ] married, spouse with me
[ ] married, spouse away
[ ] never married
[ ] divorced / widowed / separated
If you marked either 'never married' or 'divorced / widowed / separated', please go to question35.

42) Please indicate your spouse's:
a) Nationality:
[ ] Turkish
[ ] other
[ ] dual citizen (Turkish and other)

APPENDIX

43) Employment status:
[ ] not employed
[ ] employed full time
[ ] employed part time

44) a) How many of your *family* are living in Turkey? _____
***e.g., mother, father, sibling, spouse, children, or any other family member who is close to you.*

b) How many of your *close relatives* are living abroad? _____

c) How many of your *close relatives* are living in your current country of residence? _____

45) Have the 2008-2009 economic crisis in the US and the aftermath affected your views about returning to Turkey?
[ ] increased my likelihood of returning
[ ] decreased my likelihood of returning
[ ] did not change my views
[ ] not applicable

46) Please write down any comments or questions about any part of this survey. Your feedbacks would be highly appreciated.

Thank you for taking part in my survey!
Asst. Prof. Adem Yavuz Elveren
Fithcburg State University
Department of Economics, History and Political Science
aelveren@fitchburgstate.edu

Brain Drain Student Survey

Dear friend/colleague,

I am an assistant professor of economics at Fitchburg State University, MA. I have been conducting a funded research to analyze the brain drain, a very crucial issue for Turkey.

With your invaluable contribution I can comprehensively analyze the decision that Turkish students and professionals make in going abroad and returning (or not returning), obtaining important results both at the theoretical and policy-making levels.

The survey involves—in addition to information on your age, gender, and hometown—several questions on your educational background and work experience as well as on your decision to stay abroad or return to Turkey.

This survey minimizes risk for participants. It does not collect name, email address, or any sort of contact information. Data will be used in a de-identified way. The information you will be providing will be used solely for statistical/econometric analysis for a scholarly article, and the information is not going to be shared with third parties.

Data will be stored in a password-protected personal computer for three years.

The survey will take about 10–15 minutes. You will be able to skip questions that are not applicable in your case. In return, your inputs will help to better understand an important issue, brain drain, for developing countries, particularly for the case of Turkey.

You may ask any questions or make suggestions about the survey by emailing me at aelveren@fitchburgstate.edu. My mailing address is 160 Pearl Street, Miller Hall 307, Fitchburg, MA, 01420, USA, and my phone number is +1-978-665-4855.

Regarding anonymity and confidentiality or any other issues about the survey, you may also contact the Fitchburg State University Human Subjects Committee at humansubjects@fitchburgstate.edu.

I would be so happy if you let your Turkish friends know about this survey.

If you are enrolling in a foreign language program, are a student in a bachelor's, a master's, or a doctoral program, or a postdoctoral research fellow in the US, England, Canada, Australia, or New Zealand, please complete the Survey for Students.

If you are holding at least a bachelor's degree and working (academic or nonacademic) in the US, England, Canada, Australia, or New Zealand, please complete the Survey for Professionals.

GENERAL INFORMATION

Please read and answer carefully. The survey will take about 15 minutes. You will be able skip questions that are not applicable in your case.

1. Personal Information: Please indicate your
a) Gender:
[ ] Male
[ ] Female

2. Birth year: _____

3. Birthplace:
city: _____
country: _____

4. What is your current *country* of residence?
[ ] Australia
[ ] Canada
[ ] England
[ ] New Zealand
[ ] USA
[ ] other, please indicate: _____

5. How long have you been living in your current COUNTRY of residence?
Years_____ Months_____

EDUCATIONAL INFORMATION

6. What is the highest degree you hold?

[ ] high school certificate
[ ] associates degree (e.g. 2 year program)
[ ] bachelor's (BA / BS)
[ ] post baccalaureate certificate
[ ] master's degree (MA / MS / MBA)
[ ] post master's certificate
[ ] doctorate (e.g., Ph.D., Ed.D., D.Sc.)

7. In which country did you earn your highest degree?

[ ] Australia
[ ] Canada
[ ] England
[ ] New Zealand
[ ] United States
[ ] Turkey
[ ] other, please specify: _____ -

8. What is the highest degree that you plan to receive?

[ ] high school diploma
[ ] associates degree (e.g. 2 year program)
[ ] bachelor's (BA / BS)
[ ] post baccalaureate certificate
[ ] master's degree (MA / MS / MBA)
[ ] post master's certificate
[ ] doctorate (e.g., Ph.D., Ed.D., D.Sc.)

9 If the highest degree you hold is a 'high school certificate', please go on to question 12.

From which university did you earn your undergraduate (bachelor's or associate's) degree?

_____

10. What was your major? (Please indicate both if you have double majors)

_____

11. In which year did you graduate?

12. What program are you currently enrolled in abroad?

[ ] student exchange program

[ ] visiting student / scholar program (e.g., you are a TÜBA or TÜBITAK scholarship recipient enrolled in a Turkish university and completing part of your program requirements abroad)

[ ] intensive language program (as prerequisite for continuing with undergraduate or graduate studies abroad)

[ ] associate's degree program

[ ] bachelor's degree program

[ ] post baccalaureate certificate program

[ ] master's degree program

[ ] post master's certificate program

[ ] doctoral degree program, course work not yet completed

[ ] doctoral degree program, course work completed

[ ] postdoctoral fellow

[ ] other, please specify: _____

13. If you are an exchange student or a visiting student / scholar, please answer the following questions. *Others please go on to question 16.*

a) From which university will you receive your degree?

_____

14. What degree will you receive from this university?

[ ] bachelor's degree

[ ] master's degree

[ ] doctorate degree

[ ] other, please specify: _____

APPENDIX 157

15. What type of activities are you involved in at the university or research center you are currently visiting? *Please check **all** that apply.*

[ ] lab work / experiments
[ ] participating in seminars
[ ] attending courses
[ ] giving lectures
[ ] independent research activities
[ ] other, please specify: _____

16. What is your field of study?
Please be specific and indicate any areas of specialization.
General field of study: _____
Specialization 1: _____
Specialization 2: _____
Specialization 3: _____

17) If you will be receiving a degree or certificate from the institution you are currently enrolled in, please answer the following questions. *If you will not be receiving any degree from this institution, please go on to question 20.*

17. When did you start the program? *Please include any compulsory language training that formed part of the degree requirement.*
MONTH _____ YEAR _____

18. When do you expect to earn your degree?
MONTH _____ YEAR _____

19. Were you required to take part in an intensive language training program prior to being accepted into the degree program?
[ ] Yes
[ ] No
[ ] I am currently enrolled in a language program.
[ ] not applicable

20. Which of the following factors played a significant part in your decision to choose your current university/research center for studying abroad. *Please check all that apply.*

[ ] A. provided the most relevant program for my field of specialization
[ ] B. provided the best scholarship or financial support
[ ] C. having Turkish contacts at the institution
[ ] D. recommended by advisor or other professors
[ ] E. greater job opportunities
[ ] F. being with or near spouse
[ ] G. other, please specify: _____

21. Which was the most important factor? _____

22. Which source(s) of financial support do you or (did you) have available to you for your current studies abroad? *Please check all that apply.*

[ ] savings or support from family
[ ] part-time job (university)
[ ] part-time job (private sector)
[ ] part-time job (public sector)
[ ] teaching or research assistant salary
[ ] YÖK (Yüksek Öğrenim Kurumu) scholarship
[ ] MEB (Milli Eğitim Bakanlığı) scholarship
[ ] TÜBA or TÜBITAK scholarship
[ ] other national scholarship or support (including private sector)
[ ] financial support from current university
[ ] Fulbright scholarship
[ ] other international scholarship or support
[ ] other, please specify: _____

23. **Do you plan to go on to the next level of studies immediately after receiving your degree or certificate?** *i.e., continue with the master's program after receiving your bachelor's degree, or go on to do a postdoc after receiving your Ph.D., etc.*
[ ] Yes
[ ] No
[ ] I am not sure
[ ] not applicable

24. **If yes, in which city / country are you most likely to continue your studies?**
CITY: _____ COUNTRY: _____

----

25. **Which of the following factors were important in helping you adjust to life abroad?** *Please check **all** that apply.*
[ ] A. having previous experience abroad
[ ] B. the passage of time
[ ] C. support from the Turkish Student Association (TSA) at my institution
[ ] D. having spouse or other loved one with me
[ ] E. having cultural attaché/embassy support
[ ] F. having Turkish friends/colleagues at my university/college/research center
[ ] G. existence of a large Turkish community in my city
[ ] H. being able to share experiences, ask for advise via Turkish internet network
[ ] I. other, please specify: _____

26. **Which has been the most important factor in helping you adjust?**
_____

27. **Did you have any study, work, travel or other experience outside Turkey prior to coming to your current country of residence?**
[ ] yes
[ ] no
*If you have no prior experience abroad then go on to **question**.*

28. What kind of previous experience did you have abroad?
Please select all that apply.

[ ] study

[ ] work

[ ] travel

[ ] other, please specify: _____

29. What is the longest period you have spent outside of Turkey, not including your current stay abroad? _____

JOB SEARCH / WORK RELATED INFORMATION

30. In which country do you think you will be working immediately after completing your studies?

[ ] Turkey

[ ] USA

[ ] another country, please specify: _____

[ ] I do not plan to work

*If you do not plan to work, please go to **question 35**.*

31. During your current stay abroad did you apply for any jobs in firms / organizations in Turkey or in other countries?

[ ] Yes

[ ] No

*If you did not apply for any jobs, please go to **question 35**.*

32. In which countries are the firms and organizations that you applied to for jobs located? *Please select **all** that apply.*

[ ] Turkey

[ ] Australia

[ ] Canada

[ ] England

[ ] New Zealand

[ ] United States

[ ] other, please specify: _____

APPENDIX 161

33. What were your reasons for applying?
[ ] To find a full time job that is directly related to my career or education
[ ] To find a full time job (which may not be related directly to my education) after I graduate
[ ] To find a part time job to cover my education or other expenses (e.g., university bookstore, library, shop)
[ ] To make extra money during the summer months
[ ] To gain work experience in my field during the summer months
[ ] other, please specify: _____

34. During your current stay abroad did you receive any job offers from firms / organizations in Turkey or other countries?
[ ] Yes
[ ] No
*If not, please go on to **question 35**.*

QUESTIONS RELATING TO THE DECISIONS TO LEAVE, STAY AND RETURN

35. What were your main reasons for going to the country you are currently staying?
*Please mark **all** that apply.*
[ ] A. To learn a new language / improve language skills
[ ] B. In need of change / want to experience a new culture
[ ] C. Education or experience in another country is required by employers in Turkey
[ ] D. Could not find a job in Turkey
[ ] E. No program in my specialization in Turkey
[ ] F. Insufficient facilities, lack of necessary equipment to carry out research in Turkey
[ ] G. In order to take advantage of the prestige and advantages associated with study abroad
[ ] H. Preference for the lifestyle in my current country of residence
[ ] I. To be with spouse/loved one
[ ] J. To provide a better environment for children
[ ] K. To get away from the political environment in Turkey
[ ] L. other, please specify: _____

36. Which of the above was the most important reason?

_____

37. Do you think your family in Turkey would support (or supports) your decision to settle permanently outside Turkey?
[ ] They would definitely support me.
[ ] They would most likely support me.
[ ] Some family members would support me, others would not.
[ ] They are not likely to be very supportive.
[ ] They would actively discourage me.

38. Before you left Turkey, what were your thoughts about returning?
[ ] I thought that I would definitely return.
[ ] I was undecided about returning; I would wait and see.
[ ] I did not think that I would return.

39. What are your thoughts about returning to Turkey now?
[ ] I will return as soon as possible <u>without completing</u> my studies.
[ ] I will return immediately after completing my studies.
[ ] I will definitely return but not soon after completing my studies.
[ ] I will probably return.
[ ] I don't think that I will be returning.
[ ] I will definitely not return.

*If you marked one of the last two options ('not return') please go to **question 42**.*

40 a) What are your main reasons for returning to Turkey?

Please mark all that apply.

[ ] to complete compulsory military service
[ ] to complete university service (e.g., MEB, YÖK, TÜBITAK scholarship recipients)
[ ] I will return when my permitted time for working abroad ends (e.g. I am a visiting scholar)
[ ] I miss my family in Turkey
[ ] I want my children to continue their education in Turkey
[ ] after achieving specific goals (gaining work experience, completing research project) I want to apply what I have learned in Turkey
[ ] I will return after reaching my savings goal
[ ] I will return after reaching my career goal
[ ] I received a job offer from a firm or institution in Turkey
[ ] I don't feel safe in my current environment
[ ] other, please specify: _____

41) After you return, do you plan to go abroad again to settle/work?
[ ] No
[ ] Yes
[ ] Maybe

------

42. In general, how do you compare your life in your current country of residence with your life in Turkey?

42) work environment (e.g. your job satisfaction):
[ ] much better
[ ] better
[ ] neither better nor worse
[ ] worse
[ ] much worse
[ ] not applicable

43) social aspects (e.g. friendships, social relations):
[ ] much better
[ ] better
[ ] neither better nor worse
[ ] worse
[ ] much worse

44) standard of living:
[ ] much better
[ ] better
[ ] neither better or worse
[ ] worse
[ ] much worse

45) What are the main difficulties that you have faced / are facing living in your current country of residence? *Please mark **all** that apply.*
[ ] A. Being away from family
[ ] B. Children growing up in a different culture
[ ] C. Loneliness, not being able to adjust
[ ] D. Fast-paced life
[ ] E. Little or no leisure time
[ ] F. Unemployment
[ ] G. No jobs in my area of specialty
[ ] H. Discrimination against foreigners
[ ] I. Lower income compared to the income I had in Turkey
[ ] J. Higher taxes
[ ] K. Crime, lack of personal security
[ ] L. High cost of living
[ ] M. Other, please specify: _____

46. Which of the above factors are the most difficult for you?
_____

47. What are the greatest difficulties RELATING TO TURKEY that may cause you NOT to return? Please indicate how important for you the following factors are in this decision.

Please answer even if you have indicated that you will definitely return.

A. Low income in my occupation ___ ___ ___ ___ ___

B. Little opportunity for advancement ___ ___ ___ ___ ___ in my occupation

C. Limited job opportunities in my field of ___ ___ ___ ___ ___ expertise

D. No opportunity for advanced ___ ___ ___ ___ ___ training in my field

E. Being far from important research ___ ___ ___ ___ ___ centers and as a result from new advances

F. Lack of financial resources and ___ ___ ___ ___ ___ opportunities to start up my business

G. Less than satisfying social and ___ ___ ___ ___ ___ cultural life

H. Bureaucracy, inefficiencies in ___ ___ ___ ___ ___ organizations

I. Political pressures, discord ___ ___ ___ ___ ___

J. Lack of social security ___ ___ ___ ___ ___

K. Economic instability, uncertainty ___ ___ ___ ___ ___

L. Other reason, please indicate below: ___ ___ ___ ___ ___

_____

48. Please indicate the relative importance FOR YOU of each of the following factors relating to your CURRENT COUNTRY OF RESIDENCE in deciding not to return or postpone returning to Turkey.

Please answer even if you have indicated that you will definitely return.

A. Higher salary or wage ___ ___ ___ ___ ___

B. Greater opportunity to advance ___ ___ ___ ___ ___
in profession

C. Better work environment ___ ___ ___ ___ ___
(flexible work hours, relaxed setting, etc.)

D. Greater job availability in my area ___ ___ ___ ___ ___
of specialization

E. Greater opportunity for further ___ ___ ___ ___ ___
development in area of specialty

F. A more organized and ordered life ___ ___ ___ ___ ___
in general

G. More satisfying social and cultural life ___ ___ ___ ___ ___

H. Proximity to important research ___ ___ ___ ___ ___
and innovation centers

I. Spouse's preference to stay or ___ ___ ___ ___ ___
spouse's job being in current country

J. Better educational opportunities for children / ___ ___ ___ ___ ___
want children to continue their education

K. Need to finish or continue with current project ___ ___ ___ ___ ___

L. Other reason, please specify below: ___ ___ ___ ___ ___

_____

49. **If your scholarship did not have any compulsory service (i.e. serving in Turkey or repayment) would you consider continuing to live abroad after graduation**
[ ] Yes
[ ] No
[ ] Maybe
[ ] Not applicable

OTHER INFORMATION

50. **Please indicate your marital status:**
[ ] married, spouse with me
[ ] married, spouse away
[ ] never married
[ ] divorced / widowed / separated

If you marked either 'never married' or 'divorced / widowed / separated', please go to question 53.

51. **Please indicate your spouse's:**
a) Nationality:
[ ] Turkish
[ ] other
[ ] dual citizen (Turkish and other)

52) **Employment status:**
[ ] not employed
[ ] employed full time
[ ] employed part time

53. Indicate the number of children living with you as part of your family in the following age categories.

under 2 years _____
between 2-5 years _____
between 6-11 years _____
between 12-17 years _____
18 and over _____

54. a) How many of your *family*** are living in Turkey? _____

**e.g., mother, father, sibling, spouse, children, or any other family member who is close to you.

b) How many of your *close relatives* are living abroad? _____

c) How many of your *close relatives* are living in your current country of residence? _____

55. Have the 2008-2009 economic crisis in the US and the aftermath affected your views about returning to Turkey?

[ ] increased my likelihood of returning
[ ] decreased my likelihood of returning
[ ] did not change my views
[ ] not applicable

56. How did you find the length of this survey?

[ ] too long
[ ] too short
[ ] just right

57. Please write down any comments or questions about any part of this survey. Your feedbacks would be highly appreciated.

Thank you for taking part in my survey!
Asst. Prof. Adem Yavuz Elveren
Fithcburg State University
Department of Economics, History and Political Science
aelveren@fitchburgstate.edus

Index[1]

A
AKP, *see* Justice and Development Party
Anatolian, 114, 115
Arab, 1, 25, 115
Atatürk, Mustafa Kemal, 109, 120, 123
Austria, 24, 25
Authoritarianism, 1, 52, 109, 114–116

B
Baran, Paul, 9
Belgium, 25
Boratav, Korkut, 111, 112
Bozkuş, Süleyman Cihan, 124n3, 125n4
Buğra, Ayşe, 110, 113, 117

C
Canada, 24, 65, 138, 153, 154
Capital inflows, 111
CCT, *see* Conditional cash transfer
Charity, 110, 116–119
China, 16, 23, 48, 49
Civil Code, the, 122
Civil society, 111
Clientelism, 117
Conditional cash transfers (CCT), 117–119, 123
Conservative democratic, 110, 113, 114

D
Dedeoğlu, Saniye, 24, 25, 110, 124n3
Dependency Theory, 9
Development Plans, 24–27
Docquier, Frederic, 7, 10, 11, 13, 15, 17
Domestic capital, 112
Double movement, 110

E
Erdoğan, Recep Tayyip, 1, 54n5, 115, 116, 125n7
Europe, 3, 8, 17, 23, 24, 40, 41, 50

[1] Note: Page numbers followed by 'n' refer to notes.

European, 24, 25, 37, 41, 49–51
European Union (EU), 4, 24, 36, 49–51, 115, 121, 122

F
Faith-based, 117
Familism, 118
Fertility rate, 15
Fiscal deficits, 111
France, 24, 25
Frank, Andre Gunder, 9

G
Galbraith, James K., 111, 112, 116
Gender equality, 122, 124, 124n3, 134
Gender inequality, 15–18, 121–124
General Health Insurance (GHI), 116
Germany, 3, 24, 25, 35, 53n2, 54n4
Gezi Park, the, 114
Ghana, 16
Güngör, Nil Demet, 4, 14, 16, 17, 28, 32, 43–48, 54n5, 71, 105, 107, 107n1

H
Harris, John, 8
Highly skilled, 2, 10, 12–16, 24, 27, 29, 37, 132

I
Imams and preachers, 113, 119
Income inequality, 12, 111, 113
India, 23, 24, 49
Inflation, 111, 112, 115
Informal sectors, 112
Iran, 16
Islamic, 1, 2, 16, 107, 109, 110, 113–120

Islamist, 1, 114, 115
Islamization, 111, 113, 114, 119–120

J
Japan, 16, 50, 122
Justice and Development Party (AKP), 1, 2, 52, 109–120, 122–124

K
Kemalist ideology, 1, 115
Keyder, Çağlar, 113
Korea, 49, 122
Kurdish problem, the, 110, 112–115, 118

L
Labor force participation (LFP), 15, 121, 124n3
Labor Law, the, 122
Labor union, 111
Lithuania, 17

M
Manufacturing, 112
Middle East and North Africa, 25
Military coup, 27, 111, 113
Military government, 111
Minorities, 2
Morocco, 24
Muslim, 1, 115

N
Neo-classical economic theory, 8
Neo-classical migration theory, the, 8
Neo-liberal model, 111
Netherlands, 24, 25

New economics of labor migration, 8
Nigeria, 17

O
Open anocracy, 121
Organisation for Economic Co-operation and Development (OECD), 15, 16, 49, 121, 122, 133, 134

P
Penal Code, the, 122
Pious Predator State, the, 111, 116
Polanyi, Karl, 110
Political Islam, 110, 112–114
Political migration, 27
Populist, 111
Poverty, 9, 12, 110, 111, 113, 117, 118
Predator State, The, 116
Presidency of Religious Affairs, the, 115
Pronatalist biopolitics, 118
Public procurement, 116, 117
Push-pull model, 9–10

R
Real wages, 111, 112
Referendum, 115, 121
Remittances, 8, 13, 18n1, 26
Reverse brain drain, 4, 5, 24, 49–52, 133

S
Secular, 1, 2, 16, 113–115
Segregation, 122, 123
Skeldon, Ronald, 8
Social assistance program, 118
Social exclusion, 117

Stark, Oded, 8
Sub-Saharan Africa, 17

T
Tansel, Aysıt, 4, 16, 32, 43–48, 54n5, 71
Terrorism, 113
Theology, 119
Todaro, Michael, 8
Toksöz, Gülay, 15, 16, 18n2, 24, 25, 54n3, 105, 121–124, 125n5
Trade liberalization, 111

U
UK, 35, 65
Unemployment, 26, 29, 36, 44, 47, 112, 121
Unionization, 112
United States (US), 3, 4, 10, 16, 17, 23, 24, 29, 35–42, 44, 46, 50–53, 53n2, 54n6, 54n8, 65, 70, 132, 138, 153, 154

V
Violence against women, 124
Virtue Party, the (Fazilet Partisi), 114

W
Wage inequality, 112
Welfare Party, the (Refah Partisi), 113, 114
West, the, 1, 115, 118, 119, 134
Women's equality, 109
Women's rights, 16

Y
Yeldan, Erinç, 111, 112

CPSIA information can be obtained
at www.ICGtesting.com
Printed in the USA
BVHW02*0006260518
517471BV00010B/40/P